W9-BZE-201

The
Mind
and the
Moon

ALSO BY DANIEL BERGNER

Moments of Favor (a novel)

God of the Rodeo:
The Quest for Redemption in Louisiana's
Angola Prison

In the Land of Magic Soldiers:
A Story of White and Black in West Africa

The Other Side of Desire:
Four Journeys into the Far Realms of
Lust and Longing

What Do Women Want?:
Adventures in the Science of Female Desire

Sing for Your Life:
A Story of Race, Music, and Family

The Mind and the Moon

MY BROTHER'S STORY,
THE SCIENCE OF OUR BRAINS,
AND THE SEARCH FOR OUR PSYCHES

Daniel Bergner

An Imprint of HarperCollins*Publishers*

THE MIND AND THE MOON. Copyright © 2022 by Daniel Bergner. All rights reserved. Printed in the United States of America. No part of this book may be used or reproduced in any manner whatsoever without written permission except in the case of brief quotations embodied in critical articles and reviews. For information, address HarperCollins Publishers, 195 Broadway, New York, NY 10007.

HarperCollins books may be purchased for educational, business, or sales promotional use. For information, please email the Special Markets Department at SPsales@harpercollins.com.

Ecco® and HarperCollins® are trademarks of HarperCollins Publishers.

First Ecco hardcover published 2022

FIRST EDITION

Designed by Paula Russell Szafranski

Library of Congress Cataloging-in-Publication Data has been applied for.

ISBN 978-0-06-300489-4

22 23 24 25 26   LSC   10 9 8 7 6 5 4 3 2 1

*The
Mind
and the
Moon*

Before he was put on a locked ward, my brother danced on ferry decks. The ferries ran from Seattle to the islands of Puget Sound. This was back then; Seattle wasn't yet Seattle. The boats were battered and sparsely used, and one day my brother, riding across the sound, wearing a khaki Army jumpsuit and little black dance shoes, was infused by the cadence of the water striking the hull and by the engine's varied, rhythmic vibrations rising up from the bowels and traveling along the deck where he stood.

The jumpsuit buttoned up the front. He'd bought it at an Army-Navy store downtown, because he wanted a garment that would unify rather than divide his body. The dance shoes had a more basic explanation. In addition to playing the piano with promise, he had, at around the age of twenty-one, ambitions as a dancer and choreographer. How I envied those shoes! Streamlined, nimble, supple as slippers but with minimalist laces, sharply pointed, scuffed yet still a rich black, they were beautiful and full of artistic attitude.

The impact of the rough water against the bow created a steady, emphatic beat, and above that the engine delivered not only a churning rhythm but something bordering on a melody, deep and ancient, like a Gregorian chant. It was a small part of my brother's gift that he both heard, at swelling intensity, this music of water and machinery and allowed himself to be inspired and electrified by it. His body responded with a physical, visceral version of a child's wonder as she holds a conch shell to her ear and listens to its elemental communications for the first time.

He stood on the lowest deck, near the front of the cars and the slung chains, as the boat's combination of Gregorian choir and pounding drum surged through him. He lifted one foot to knee height, then leapt high off the other and landed on the first foot, so that there was a simultaneous vaulting and transferring of weight, followed by a reversal and more repetition back and forth, melded with the strivings of his torso and arms, amounting to movements at once airborne and sinuous. To the few passengers who watched from their cars, his mix of military jumpsuit and elfin shoes may have looked odd, compounding the oddity of his dancing, but all of this strangeness was countered by the broad solidity of his body and by his resistance to the sporadic lurching of the boat, which should have pitched him off balance and made him grab at the chain poles or brace himself against a car, but never did. He hung in the air, stomped his heels on the steel deck, sprang from side to side, spun, and elevated again, athletic, animalistic, ethereal, impelled by the pulse of the water and the echoes of medieval worship.

And soon he was on a psychiatric ward, with a heavy dose of Haldol seeping into his brain.

This is his book. And it is the story of a few of the many who, over the past several years, his story sent me out to find.

ONE

Caroline was living in a group home when her picture went up on billboards around Asheville, North Carolina. She floated above the Bojangles Restaurant on Merrimon Avenue. She hovered over Tunnel Road and above the freeway running out of the city toward Hendersonville. In the giant picture, she crouched, poised for a rush of speed. Her jammer helmet sat low over her dark eyes. Asheville is a small city near the Tennessee border, but it's the heart of a metropolitan area with half a million people, so a great many drivers passed beneath her image.

It was strange for her to consider, this abrupt bit of fame. But suddenly she was a star on the city's flat-track roller derby team, and in the packed Civic Center mothers leaned at the rail with programs for her to autograph and gratitude to proclaim. They said they wanted their daughters to be just like her. They seemed to perceive, in her pint-sized being, an outsized strength. This made everything all the more bewildering, because she, who was scribbling her name across her photo in the programs, who was glancing up at the billboards from the bicycle that got her from here to there, or looking up from the back of the group home van that took the residents to the appointments and volunteer jobs that were supposed to scaffold their lives, or later gazing up from the old sedan her uncle gave her, was perpetually hearing voices, hearing people who weren't in her car or in the group home, weren't standing in front of her or looming behind her shoulder, who didn't exist, not in the way most people define existence, but who were perfectly real to her, yelling, whispering, wailing, warning, commanding, beseeching, berating, and, not infrequently, instructing her to kill herself. She was in her late twenties. Abilify, Risperdal, Depakote, lithium, Seroquel. The staff at the group home believed she was still taking her medications.

She had grown up in Zionsville, outside Indianapolis. There, before school, she had breakfast in a bright kitchen whose wallpaper was harvest gold. The linoleum tiles had a cobblestone motif. A bowl of oatmeal was set out for her on the round table by the broad window. Beside the white bowl was a sun-speared glass of water and a white saucer with an array of pills. Oblong, jumbo, and pink; orange and petite; half yellow and half green—the shape, size, and color of the pills varied over the years, but there was always a combination on the saucer, never just one, not by the time she was in middle school. As a middle schooler, she didn't yet have a clear knowledge of their separate purposes. They would help, she was told, and had been told since she was eight or nine.

Earlier, back when she was in daycare, one of the daycare ladies had remarked on the beautiful weather. "Doreen," she said to her co-worker, "there isn't a cloud in the sky today, not a single cloud." Caroline was sitting next to a Big Bird Big Wheel. A voice told her that there were, in fact, clouds in the sky at that moment. "Those ladies," he said vehemently, "are liars."

At home, she started to play Sorry with this voice, moving his pieces. During elementary school, he warned that her father was at risk of dying, her uncle as well, that her entire family and people beyond, Zionsville in general, were in unspecified peril. This danger was tied up with the Gulf War, with the TV news images of night-time bombing that followed her to bed: fighter planes, flashes in the sky, explosions on the ground, luminous and all-consuming. The voice never clarified the link between the war and the imminent family threat, but he communicated, constantly, the danger and a need to intervene—intervention he could not carry out on his own.

She shared with her parents part of what she heard and part of what she sometimes felt, a tugging at her arms, her neck. It was around then that medication was introduced to her morning routine. Nevertheless, new voices rose up. Unlike the first, who mostly felt like an ally, these new additions belittled her and derided one another. They could grow loud enough that it was hard to catch the words and make meaning of the sentences spoken by those whom others did perceive.

Bodily changes occurred as medication was adjusted, added. She watched herself develop, in middle school, a coating of lard. Then she watched herself become obese. And she was losing control of her forearms and hands. They trembled; they shook. Her hands seemed to want to flap-paddle the air.

About her size, she absorbed the lesson that everyone else seemed to agree on: she was lazy; she gorged. She found some old exercise videos and performed exactly as many crunches as the teacher in the belted leotard; she got down on all fours and lifted her legs to the side just as many times as the teacher lifted hers; she stepped, stepped, stepped in sync with the teacher's aerobic example. She did this not once or for a week but over a period of protracted faith, and she mixed this regime with a pared-down diet. It accomplished nothing. So, in school, she was an obese girl whose pencil quivered on the page as she took notes in social studies or tried to show her work neatly in math. She was also losing hair, wrapping several strands around her fingers and yanking them out, twisting and pulling consciously and unconsciously, needing to, compelled to. Patches of bare scalp crept into view.

"Are you a crackhead?" the girls of Zionsville Middle School asked.

The school's low building sprawled, bunker-like, surrounded by odd grassy mounds, making her feel that all within was concealed.

"Where's your pipe?" the girls asked along the locker-lined halls.

They had straight, thick, gleaming hair, parted down the middle. They wore heavy eyeliner and scant baby doll tops, and they shoved her into the lockers' metal doors.

"What a fucking fatass."

Behind her, a girl put a hand at the base of her head and rammed her forehead against the steel.

"What a fucking freak."

In class, in the cafeteria, in their impossibly white tee shirts with tiny capped sleeves:

"Why is your hand shaking?"

"Why are you fucking bald?"

"Why are you fucking crying?"

"What is fucking wrong with you?"

She wondered if she might be a different species, another genus. Genetically alien, she found some refuge within the grunge fashions that had made their way from Seattle to Zionsville. Kurt Cobain's slack, stained, nubby cardigans inspired her; she hid her body in the disintegrating button-down sweaters she took from her grandfather's drawers. She wore baggy brown polyester trousers from her father's closet.

But her main refuge was in books, not only because they helped her to forget the girls and flee the thoughts of her own extreme evolutionary difference, but also because the cacophony of her voices could be overridden or quieted, to a degree, by reading—

especially by reading aloud, to herself or her two little sisters. For herself, she chose almost randomly from her father's and grandfather's shelves. She read *Ethan Frome* over and over. She raced to the end, to the suicide attempt amid the snow and stars: the impulsive resolve, the expectation of escape, and then what Edith Wharton called the "delirious descent." That the attempt went horribly wrong, that the descent ended not in death but in permanent crippling, hardly mattered; it was the plunge, the feeling of flight, the release, the delirium that left the most powerful impression and led her to reread and reread.

Soon, early in high school, there was *Hamlet*: "O . . . that the Everlasting had not fix'd / His canon 'gainst self-slaughter!" And she chanced on Alexander Solzhenitsyn and his most famous character, Ivan. Ivan's eyes, directed distantly elsewhere, stared out from the tattered paperback cover. He was serving three thousand, six hundred and fifty-three days in a Soviet labor camp, in temperatures of thirty or forty below zero, for a crime, spying for the enemy, that he'd been forced to confess but didn't commit. He was shedding hair and teeth and wondering how, if he survived, he could ever explain his existence in this frigid prison to anyone who hadn't experienced it, anyone on the infinitely far side of the divide between what he knew and what was known by others. He put the problem simply to himself: "How can you expect a man who is warm to understand a man who is freezing?"

DAVID WANTED ME to hear him play the ukulele. "Wait," he said, and went to get it. This was at the initial height of the pandemic; we were on the phone. It was dawn, and he was in the detached studio he and his wife had added behind their house. It was decorated with deco lamps and old film posters from Latin America, and was supposed to be a place for the family to watch movies or spend time with friends. But it had become the cell where

David, to avoid disturbing his wife, endured his insomniac nights. Three or four hours of sleep was typical. This had been going on for around two years. It was one of the torments, though not the worst, that he hoped somehow to cure, or even just slightly reduce, by undergoing a series of psilocybin treatments, beginning the following week. He was terrified by what he'd signed up for. The guide who would oversee his psychedelic sessions was a man named Dune.

"You ready for something really painful?" he asked me, and began a rendition of "Red River Valley." He took ukulele lessons once a week, meeting with his teacher by Zoom. He'd chosen to learn the instrument partly because his teenaged daughter had asked for one and abandoned it, so it was there for him to try, and partly because he'd decided that music lessons, on any instrument, might help him: the discipline, the distraction, and the generation, the members of his internet support groups told him, of neuroplasticity, of healthier neural pathways. He also swam serious distances, ran hills and stairs, and, for a few minutes each morning, attempted to meditate. Now he strummed the opening chords of the nineteenth-century cowboy love song, and interspersed his strumming with dapplings of plucked notes, all with an innocence that was accentuated by playing into the phone, which blurred the music gently.

David was a lawyer. The civil rights organization he worked for had long been struggling against the prison system in his southwestern state, filing complaints in court and sending out press releases about overcrowding and consistent mistreatment. The problems were covered fleetingly by the local media before being forgotten. The lawsuits achieved little. When David took over the battle, he and his team published a report containing more flagrant abuse: a dozen testimonials about inmates being brutally beaten by guards. In response, the superintendent of the system mocked David's organization as an outfit of suckers and bleeding hearts, and then the media learned that one of the testimonials came from a convicted child molester. Pedophilia replaced prison conditions as the story of the moment.

The next time, David told his colleagues, they weren't going to make a single allegation, weren't going to issue so much as a peep, until they had collected one hundred inmate statements. This seemed an absurd threshold, but one victim of assault by guards seemed to lead to five others, and one witness led to the next and the next, and as the team neared seventy, David had the realization that a group of employees just might be willing to add their testimony: prison ministers. He had no clue how to contact them covertly, but relentlessness and serendipity put him in touch with two chaplains. They knew well the violence perpetrated by prison staff, but they were afraid to speak, scared less for their safety than for their vocation. The prisons were their calling; if they were banished by the superintendent, life would lose meaning. And no matter God's omnipotence, the superintendent was all-powerful.

David found a first-rate litigator to represent one chaplain pro bono if the superintendent retaliated, and he arranged for the other to stay anonymous; gradually both men were convinced. David's team tracked down more non-inmate witnesses. Two prison experts reviewed the evidence and gave damning assessments. David made sure that national papers and networks paid attention. Guards had choked an inmate until he lay unconscious, then kicked him again and again in the head. They had sodomized a convict with a flashlight. They had broken jaws and fractured facial bones. They had pepper-sprayed an inmate who was wheelchair-bound. They had cuffed a man who was visiting his brother, and lacerated and pummeled the visitor's face until his eyes swelled shut. They had shattered a convict's eye socket, ordered him to strip and walk naked along a corridor, and then locked him into a cell, where gang members took turns raping him. They had unlocked a cell door, leaving an inmate to be sodomized by other convicts with a broomstick. They had etched a racial slur into an inmate's skin. They had formed a gang among themselves, with initiation rites requiring attacks on inmates and fellow officers.

The accounts led to public hearings. This, in turn, kept the atrocities in the media. David brought a new lawsuit, and the presiding judge admonished the department of corrections that its defense was untenable. The department submitted to federal oversight. Guards were convicted of assault and obstructing justice. The superintendent was forced to resign and found himself locked up in a federal prison.

But this triumph felt, to David, relegated to a remote past. It had culminated several years ago; it seemed that he was not the same person now. When he contemplated the memory of that crusading lawyer and compared it to the attorney he had lately become, he could barely connect the two. There was scarcely a resemblance. He was still with the same organization, and his name appeared on legal briefs arguing for the right of protesters against police racism to march after dark, and for the need, during the pandemic, to spare the lives of nonviolent inmates by freeing them from densely populated prisons and jails. His name was prominent on the paperwork, but his mind was nowhere in the words. The credit he was given was a charade, he felt, a gesture of charity, a nod to his title within the organization, to legal ingenuity he no longer possessed.

He couldn't think the way he once had. He was only fifty, but when he set out to study judicial rulings that would add gravity to a current brief, he couldn't keep the cases in mind let alone make connections between them and fit them together into a compelling theory. Worse, he had trouble not only with the theoretical but with the factual. He strained merely to tell the story of the harm a plaintiff had suffered. A coherent narrative, with well-placed, attention-getting details, was an essential part of any suit, and the skill he'd once had as a storyteller was vaporizing along with his talent as a thinker.

Days before he played "Red River Valley" over the phone, he had a mortifying experience. He was on a conference call with a director from the organization's national headquarters and some of the young lawyers on his regional team. In response to the director's questions, the young attorneys did all the talking, and the longer the call went on, the more David's own silence seemed to echo. He had nothing to contribute. His colleagues had done most of the work—and come up with all of the conceptual architecture— on the memo they were discussing. Hours earlier, his eyes had moved over the memo without even basic comprehension. He wondered how many more minutes would pass before the director would ask for his input; he knew the director must be keenly aware that he was speechless, mute, dumb to an extent that couldn't be explained by letting his junior teammates learn through taking charge. He listened to the conversation and labeled himself cognitively impaired.

Yet more desolating than his inability was his loss of interest. So much passion had surged through him during the prison case, his devotion never wavering when denials and long delays were put in his path. A crusading confidence, an empathy for the powerless, a sense of moral crisis drove him. This had been true from the beginning of his career, and as recently as three and a half years ago, the shock of Trump's election, the ugliness of his policies and pronouncements, the revelation of the country's widespread capacity for cruelty had felt, to him, like a physical wound. Somehow, though, this gave way to mounting indifference. *Fuck it*, he often thought nowadays. *Who cares? I don't care about this at fucking all.*

His internet support group friends supplied him with an explanation that he did and didn't accept. Around two years ago, he'd decided to find out who he was without psychotropic medication, and they were certain that his withdrawal from an antidepressant and then from an antianxiety drug, a benzodiazepine, had created his torturous situation: his unremitting insomnia; the sluggishness of his mind, a torpor that seemed too heavy to be attributed to his sleeplessness alone; his disengagement not just from the mission of his work but from his wife and only child; his fathomless despair; the self-absorption that felt like despair's Siamese twin; the unshakable thought that his wife was married to a cripple; and the nerve pain that came quickly after his intervals of sleep and that persisted into every morning—a sensation of drastic heat, of being scorched, that overtook large sections of his body, ankles, shins, thighs, stomach, or that struck almost everywhere, a feeling that infinite individual nerve endings were aflame, from feet to forearms to shoulder blades to neck to forehead to ears, or, on especially bad days, that seemed also to set his lips and tongue on fire.

Cold packs and cold showers couldn't do much to relieve the scorching; each morning, he endured it until it dissipated. But as excruciating as it was, he was sure that if some cosmic force or deity he didn't believe in offered him the deal of removing his despair and detachment while leaving him with his nerve pain for the rest of his life, he would eagerly take those terms.

Dune, the self-proclaimed guide for psilocybin treatments, predicted an improvement in all his symptoms. He had been recommended to David by a relative. David had only spoken to Dune over the phone before booking the first of three sessions for next week. He recognized the irony of wanting to free himself from psychotropic medications yet meanwhile resorting to another chemical, but he took solace because psilocybin was natural and he would only use it three times.

He had done some research. He wasn't going by Dune's claims. There was a neuroscientist, Roland Griffiths, a professor at Johns Hopkins, one of the most prestigious universities in the country, who had been researching the effects of psilocybin for almost two decades. David had watched him online. A tall, wispy man with a full head of gray hair, Griffiths wore a navy blazer and an open-collared shirt and began by saying, in a faintly grainy, soothing voice, "Most people assume that science and spirituality don't play well together." But Einstein, he continued, considered "the mystical" to be "the source of all true science."

Psilocybin, Griffiths said in a relaxed tone, as if his emerging data put him beyond self-doubt, engenders "mystical-type experiences" that lead to "enduring positive changes in mood." On the screen behind him, he showed a picture of the couch and tribal art in his low-lit session room at Hopkins. Through one or two hallucinogenic journeys, his research subjects gained a long-lasting "sense of unity, a feeling that all people and things are connected . . . joy." He projected statistics onto the screen. For ninety percent, there was "increased life satisfaction" and "better social relationships," outcomes that were corroborated by interviews with family, friends, colleagues. In one study, his subjects had life-threatening cancers and suffered from clinical depression and anxiety. After treatment, eighty percent showed "substantial and enduring antidepressant and anxiolytic effects." For most, despite their physical illnesses, their psychological symptoms faded into normal range.

David saw himself as rational to a fault, and Griffiths's talk of Einstein and merging the spiritual and scientific worried him. Was he too much of a skeptic to be susceptible to psilocybin's curative influence? This question was troubling enough; more worrisome was whether, for him, the treatment was highly dangerous. He imagined two risk factors converging. First, he was a control freak. He'd smoked marijuana all of twice in his life, and as for any other illicit substances that might jeopardize his self-control, he'd avoided them entirely. Could this caution be exactly what his particular brain needed? If he shut out his fear and swallowed a psychedelic, wasn't he beckoning insanity?

The second factor concerned his brain in its current state of medication withdrawal. His circuitry was damaged and destabilized. His GABA receptors, he'd read, were decimated. His weakened brain was struggling to heal. How could it possibly be safe to take a psychedelic in his condition? Wouldn't the chemical ravage him? Couldn't he be left with a mind beyond recovery? Wasn't he a prime candidate to wind up crazy?

Petrified, he sent an email to Griffiths, noting his withdrawal from psychotropics, his trepidation. The scientist was kind enough to reply. He said that he saw no reason, in the situation David described, not to proceed. David wrote to Dune. The guide said that not a single client of his had ever had a bad journey.

This morning, with his appointment four days away, David played the ukulele into the phone. "I'm not musical," he told me before he began. "But I love music, and I've learned 'Red River Valley' and 'Song of the Volga Boatmen.'" He staggered through "Red River Valley." Each strummed chord and each plucked note was delayed; they bore only tenuous relationships to each other. The tune was only dimly recognizable. "No," he interrupted himself. "That was the wrong chord. That was terrible. Let me play it one more time." He started again from the first measure, as hopeful and exposed as a child.

I WOULD NEVER have met Caroline or David—or Chacku or the others with our conditions—were it not for my brother.

Were it not for my younger brother, Bob, I would never have met the scientists who study our brains with the ultimate hope not only of treating our conditions but of understanding our minds, of crossing the chasm between physiology and consciousness, between circuits and thought, between cells and emotion, between matter and self—between what we're made of and who we are.

Were it not for my brother, I would never have measured that chasm against the distance between here and the moon.

Were it not for my brother, I would never have spent time with the neuroscientist and former director of the National Institute of Mental Health, the NIMH, the largest mental health research organization in the world, who viewed the brain through the lens of a short story by the Argentinian postmodernist Jorge Louis Borges, "The Garden of Forking Paths," a tale of labyrinths and innumerable bifurcations, and through the ancient myth of Theseus and the Minotaur.

Were it not for my brother, I would never have met the psychiatrist who had renounced a career of conventional success to open homes for those with psychosis, with hallucinations and delusions and the most extreme diagnoses—homes whose curative treatise was a rabbinic parable, "The Turkey Prince."

Were it not for my brother, I would never have found myself underlining sentences in a 2019 lead article in the *New England Journal of Medicine* examining "psychiatry's identity crisis."

Were it not for my brother, I would never have confronted all that we do not control—or, despite confronting this constantly, as so many of us do, I would never have contemplated it the way I have: wondering about the cost of our belief in biological psychiatry; calculating and recalculating the cost of our centuries-old faith—rooted in the Enlightenment—in the potential of science and medicine to right our minds; tallying and retallying the price we pay in our desire to keep ourselves and those we love away from life's agonies and clear of its precipices.

And to tell all I need to about my brother, I need to say something about our parents.

Our father was a frightened man. This wasn't easy to detect. As a physician and epidemiologist, and as a public health official, he projected a quiet authority and had a talent for keeping people safe. In New York City, in the late 1960s, he led research into an epidemic of children falling from apartment windows, tumbling to their deaths or suffering terrible injuries. This was happening most often in poor neighborhoods in the summer months. Air conditioners were scarce; windows were wide open. In the South Bronx, one policeman told investigators that in a single summer he'd picked up the bodies of nine children who'd died in this way. Our father helped to create a city program that installed free window guards in the homes of struggling families and included a campaign to raise awareness. The program was called Children Can't Fly, and it reduced the problem by half.

There is a photograph of our father taken around that time. He is being sworn in by the city's mayor, John Lindsay, as an assistant commissioner of the health department. Our mother is in the picture, as are my brother and I. Handsome and heavily decorated for his feats in the Navy in World War II, Lindsay was compared, back then, to JFK. In the photo, he holds a Bible and smiles down at our father, who lifts his right hand. He is elevating our father into the upper ranks of his progressive crusade, anointing him to use medicine and research, reason and common sense, to protect New Yorkers from harm.

My six-year-old brother stands in front of our father in the center of the picture, his minimal shoulders encompassed by our father's body, his head on a level with our father's lower ribs. He's dressed for the event, all three buttons of his herringbone blazer buttoned up. His light brown hair, with long, thick bangs, is neatly brushed. Yet he is the only one of us who seems unconnected to the moment. Our mother and father are beaming. Bob's head is tipped to the side, an earlobe peeks out from under his hair, and he gazes off at an angle, heedless of the camera, looking at once impish and submerged in sad thought.

Children Can't Fly was an experimental effort funded by the city, but it led to a law forcing all New York landlords to provide guards in any apartment where there are kids ten or younger. Soon we moved to Seattle, where our father had been named the director of the health department, and there his achievements were just as unflashy, just as reasonable, as mandating those metal bars. He defended the city's policy of putting fluoride in the water against the accusation that the chemical was a Communist plot; he argued on the radio, in his restrained, almost monotone voice, that low levels of fluoride were a way to prevent tooth decay, especially in low-income communities, where dentists were in short supply. And for heart attack victims in Seattle and the surrounding county, he made sure that 911 phone dispatchers were ready with CPR instructions when bystanders called in, and he put state-of-the-art equipment into emergency medical vehicles, equipment that could shock the heart back into beating.

Online, I recently came across a newspaper headline from a few years ago, long after our father retired. It said that Seattle was still the "best place in the world to have a heart attack," and the story traced this record back to his time. Reading this, I thought that perhaps he was a minor type of hero, though his manner was utterly undramatic and though his accomplishments were inconspicuous. He left a trail of writing in professional journals; a publication called *Injury Prevention* chose one of his articles as an "Injury Classic." Can heroism come with such dry credentials? Is such thinking of mine delusory? The remnant of a child's need? In any case, the lack of flashy achievements fit with his unspoken credo, that solid evidence, clear eyes, and deliberate, rational action could safeguard us all.

And then he faced the possibility that his son, my little brother, was severely mentally ill. He had no idea what to do, how to hold this threat at bay. Nor did our mother, a professor of medical sociology—like him, a rationalist. They decided that my brother, at twenty-one an aspiring musician and dancer, a young person who saw himself in artistic flight, should be committed to a locked

ward and to the best judgment of its psychiatric staff, who judged
that an antipsychotic should be sent into his brain.

On that ward in the early eighties, my brother was near the
forefront of trends that quickly took hold and that have kept their
hold ever since, trends coalescing around the certainty that our
psyches are best understood and, whenever necessary, whenever
beset with conditions spanning from depression to deep breaks
from shared reality, are best aided or cured through neuroscience
and biochemistry. The swift diagnosis that he was bipolar; the
major doses of Haldol, Thorazine's cousin; the amantadine to
combat Haldol's Parkinsonian side effects; the lithium he was told
he would need to take for the rest of his life, pills that left his hands
tremoring, so that playing the piano with his previous agility and
emotion became impossible—this was the product of psychiatry's
scientific advances.

When our parents arranged to have him committed, they were
too distraught to give me a coherent explanation of what was going
on. They were in a state of dread. They could only tell the story
in fragments: my brother, who had dropped out of college, empty-
ing his bank account; my brother flying from Seattle to New York,
on a mission to cure our grandfather of Alzheimer's. He believed,
our father said, that he had this power. He believed that he might
be the messiah. He had declared this, our mother said, in a letter
written backward, a letter in mirror-image handwriting. He had
been disheveled. He hadn't been sleeping. "Erratic . . . angry . . .
ecstatic"—they exhaled words into the air, then seemed bereft of
all energy.

We stood in what had been my brother's bedroom. Home
because of this emergency, I had stepped into his room, alone,
to reckon somehow with what had happened. The room was an
attachment to the rest of the house, a kind of annex. He had always been separate, and—because he was better-looking, a better
athlete, only sixteen months younger—I'd done what I could to

keep him that way. There had been a measure of violence in my methods. I stood there, staring at the single bed with the drawers beneath it, the desk at the latticed window looking out at the fir trees, the Aboriginal bird sculpture and sacred bark paintings he'd brought home, the year before, from a solo trip to the north of Australia.

I was about to bend down and hook a finger through the brass ring on one of the drawers under the bed. I was about to stare at whatever was inside: toys, forgotten clothes, hints. Then our parents had stepped into the room behind me. They were desperate to share what they were going through and have me agree with their resolve that he was ill, that the hospital was the right place for him, medication the needed remedy. The Aboriginal art, leaning and lying around us, felt like a hapless effort, by my brother, to say something to them about himself. Our parents' voices flagged; their faces were drained. The three of us fell silent until our mother said, "Give him the book."

Our father left and returned and handed me a paperback.

"We can't right now," our mother said, "but this will tell you everything."

"It's important," our father said, rallying, his voice soft but stern.

Our mother told me that the book was new, and they drifted from my brother's room. I sat down on the bed and opened to the first chapter, titled "The Third Revolution." I read: "Disorders of mood throughout the centuries have been misdiagnosed and, at the very least, unsuccessfully treated until recently. Those that have not led to suicide have often remained uncontrollable, even in the hands of experts."

I didn't pause to dwell on how our parents must have processed those lines, as a promise and a dire admonition: that my brother's condition could be treated; that if they didn't accept the advice of the author, a psychiatrist, my brother might kill himself. I read:

"What I have to say in the pages to follow may startle many who believe primarily or exclusively in psychological approaches. . . . Anyone who has kept abreast of the new chemical advances for the treatment and prevention of mood disorders, however, will know that we are now undergoing our third and most spectacular revolution in the treatment of emotional states." The first revolution, the book said, had begun in the late eighteenth century, led by a French psychiatrist who steered the profession away from chains, bloodlettings, and beatings. This humanistic breakthrough was followed, a century later, by the second revolution, Freud's construction of the unconscious and theories of repression—although, about Freud, the author made something clear: "Originally he had been a neurologist, and he never gave up the idea that all psychological illness must in the end be attributable to an organic process that would one day be treatable with chemical therapy."

The author gave himself a featured part in the third revolution, which was defined by psychiatry's entering "into the field of medicine," and which had started with discoveries in the 1950s and was now culminating with his announcement that bipolar disorder was massively underdiagnosed, that it could be chemically cured "eighty-five percent of the time" with "a simple, naturally occurring substance," a salt, lithium carbonate, and that "millions . . . could lead normal lives after years of waste and suffering." But the third revolution, "this revelation, which is now gaining acceptance in most scientific circles around the world," wasn't limited to bipolar disorder and lithium, not at all. The schizophrenic—with their more pronounced and prolonged departures from common reality—and the depressed as well could be cured by medication.

I closed the book. I took in the brightly colored back cover— "New medicines for the mind"—and went downstairs, where I heard our parents in the kitchen. Outside the kitchen-door window was an area that had once belonged to the burly, tufted, bearish German shepherd puppy we'd named Ursus. This was a reference to the Great Bear constellation in the northern sky: a leaping, playful animal made of stars. We'd given our puppy a male version of the constellation's name, Ursa. Ursus had been an exuberant experiment for our family, a first dog whom we'd all failed to take care of and had to return—with a staph infection he developed after cutting himself in the fenced dog run where we left him unattended far too much of the time. In the kitchen, after he was driven back to the kennel where we'd chosen him months earlier, our father and my brother stood in each other's arms beside the stove. I stood near the breakfast table. It was a long room. There was distance between us. I stood apart, watching them cry, feeling far less than they did, a stalactite of snot dangling from our father's nose almost all the way to the top of my brother's head and then falling into his lush, light brown hair.

They were both so fragile. I sensed this then, when I was twelve and my brother ten and a half. I don't mean that our father suffered from what may have plagued my brother, only that I felt their mutual vulnerability. They clutched each other by the stove for quite a while, seeming to merge physically, our father's rib cage open and enwrapping my brother's head.

Our father, I knew, had been deserted by his father when he was a child, and I understood enough, vaguely, palpably, to link his knowledge of the flimsiness of family with his focus on safety. Ursus had been bound up with this, a sign of our happiness and solidity. We had just moved from an apartment in Brooklyn to a house in Seattle that was surrounded by levels of lawn, trees with a heron's nest, a driveway with a basketball hoop; we had left behind a life our parents felt was fractured and harried for one that would be more whole, and even before we arrived, a dog was a character in the tale we told ourselves. But Seattle proved to be a foreign city. On some evenings, our mother lay on the living room floor and cried out for our father's aid.

Ten years later, I found our parents in the kitchen, wordlessly confronting the chance of their son's self-destruction or death. Our mother's gaze, usually full of layers, was flat, impervious, un-yielding. Our father's eyes seemed transparent; they wandered, were lowered and raised, revealing a churn of panic. He wasn't equipped for this. This was as far as possible from a problem that could be solved with window guards, or from a rescue that could be accomplished with a defibrillator, whose jolt would revive the muscle that is the heart.

TWO

E ric Nestler was searching for new ways to treat our depressions. Donald Goff was doing the same with our psychotic
conditions—the disorders diagnosed by our voices, terrors,
conversations with all manner of gods and cosmic powers, the
conditions defined by our most subjective states. Nestler and Goff
were among the most prominent neuroscientists in the country.
They were peering into subregions of the brain. They were illuminating the relationship between experience and our ever-reacting,
ever-rewiring circuitry, what is known as neuroplasticity. They
were catching glimpses of the fibers of learning and memory,
gleaning insight into the neural systems that constitute character,
the physical and functional stuff of the self.

Nestler, with his penetrating brown eyes and smooth scalp,
and Goff, with his wide blue eyes and silver hair, had been at this
since the eighties, seeking medications along with explanations for
why we are who we are. They ran brain and psychiatric research
programs, Nestler at the Icahn School of Medicine at Mount Sinai and Goff at New York University's Langone Health. Nearly

four decades of exploration hadn't drained excitement from their voices. When we first met in 2020, Nestler spoke about "falling in love with the nervous system" in college, his doctoral work with a Nobel winner, and his current research into resilience, into what separates those who seem inexplicably, enviably buoyant from the rest of us.

Then he got metaphysical, talking about apprehending consciousness and interpretation, about how we understand a tree to be a tree and springtime to be springtime—and how we get from there to joy or perhaps an undertow of sadness when we see a tree in May. Yet elusive as such comprehension was, he said, "the capacity of our brains to understand ourselves is enormous." Goff talked like this, too. "The brain is this beautiful, complex—" He cut himself off, going briefly silent, unable to find a noun that was equal to his meaning. "The fact that the brain can even begin to comprehend how the brain works is remarkable." Moments later, his mind danced along paths leading to and from the seahorse that lies in the brain's depths, the hippocampus.

But they were blunt about the failures of their calling. Nestler said it would be easy to argue that—with a couple of very imperfect exceptions—"there hasn't been a truly new mechanism for the treatment of any psychiatric disorder in over half a century." And the present options were crude in their mechanics and problematic in their results. Goff described a pattern of hypothesis and hope, followed by promising, theory-affirming data in preliminary drug trials, followed by dismal, dispiriting outcomes in later phases of testing. He laughed ruefully about the disappointments of his career, the dead-ends psychopharmacology had run into. There was an overtone of relief in his voice, the release of confession.

Their appraisals shouldn't have surprised me. Months earlier, the *New England Journal of Medicine* had declared, "Something has gone wrong in contemporary academic and clinical psychiatry. . . . We are facing the stark limitations of biological treatments. . . .

Ironically, although these limitations are widely recognized by experts in the field, the prevailing message"—to everyday practitioners and the public—"remains that the solution to psychological problems involves matching the 'right' diagnosis with the 'right' medication. Consequently, psychiatric diagnoses and medications proliferate under the banner of scientific medicine, though there is no comprehensive biologic understanding of either the causes or the treatments of psychiatric disorders."

The proliferation pointed to by the *New England Journal* meant that more than forty million American adults and millions more children and teens were on psychiatric drugs. It meant that over the past decade, the number of adults on these medications had climbed steeply. It meant that during one recent ten-year period the number of kids diagnosed as bipolar had risen forty-fold. It meant, Nestler said, that "thirty to forty percent of our college students are treated with psychiatric medication at some point in their college years—and there's no way that their incidence of mental illness approaches forty percent." And it seemed to mean something particular about the United States, where diagnoses and the use of psychotropics were more widespread than in comparable countries. ADHD diagnosis in kids was between two and ten times higher than in European nations. A 2011 study done in the United States, the UK, and Germany had adult actors feign depressive symptoms for hundreds of clinicians. It found that you were around twice as likely to be treated with medication in the United States as in the UK (though British antidepressant use may now be increasing toward U.S. rates) and that U.S. clinicians opted for medication thirty percent more frequently than German practitioners.

But the *Journal* was also calling attention to something beyond the proliferation. This was the vast divergence between, on the one hand, the ongoing rush to prescribe, along with a persistent societal faith in the chemicals offered, and, on the other, the murky benefits of the medicines, the harm they could do, and the sparse scientific knowledge behind them. So the way Nestler and Goff spoke shouldn't have surprised me. Yet their candor was striking coming from scientists who had spent their lives on quests for medications and who were still avidly committed to those quests.

"Our field desperately needs new targets in the brain," Goff said, "and novel ideas about how to proceed. New paradigms." He had novel hypotheses of his own. His mind returned to sections of the seahorse. "We desperately need a success."

Esther Blessing, one of Goff's collaborators and a young professor in his department at NYU, had shifted away from psychiatric hospital work and toward research, because of her unease over medicating patients with antipsychotics against their will, even when she deemed it the better of bad choices. She knew the side effects, the wreckage the drugs could inflict: tremendous weight gain; ceaseless tics and spasms. A body beyond self-control. The destruction of a life. In the way we treat psychosis, she said, "there's been a gradual advancement of barbarism. We're a lot better than Antonio Moniz, a lot better than the lobotomy van. But there's still a degree of barbarity."

BEFORE LOBOTOMY BECAME a cure for afflictions ranging from delusions to depression, from mania to anxiety, a century and a half before, Britain's King George III, in the late 1780s, began talking "incessantly and violently," one of his attendants wrote, "on strange varieties of subjects." He raved in "wild incoherences." He paced castle floors through sleepless nights and gave orders to hallucinations. He spoke with a long-deceased daughter. In protest against the attempts of his court to manage him, he tore off his wig and peed on himself. A royal building, he believed, had just collapsed. London was about to be flooded. He composed messages to Don Quixote.

Psychiatrists today have looked back on the historical evidence and diagnosed the king. Some have said he had bipolar disorder accompanied by psychosis; others have diagnosed him with porphyria, a rare disease of the hemoglobin that can lead to hallucinations and delusions. His physicians, in any case, were at a loss. One of them fretted that the monarch's nighttime reading of *King Lear* enflamed his malady and made him "ungovernable." Then the wife of a man who tended the queen's horses told the court about a doctor who had treated her mother's madness.

Francis Willis, a clergyman, physician, and the owner of a small private asylum in the countryside where those with means sent their family members, was called to the palace, upriver from London, where the delirious king was kept hidden. At his asylum, Willis's methods were, in some ways, progressive. He combined the preservation of dignity with the benefits of farm labor. One memoirist of the time escorted an eccentric actress to Willis for evaluation, and wrote that "as the unprepared traveler approached the town, he was astonished to find almost all the surrounding ploughmen, gardeners, threshers, thatchers, and other labourers attired in black coats, white waistcoats, black silk breeches and stockings, and the head of each 'bien poudré, frisé, et arrangé.' These were the doctor's patients; and dress, neatness of person, and exercise being the principal features of his admirable system, health and cheerfulness conjoined to aid the recovery of every sufferer attached to that most valuable asylum."

Willis was emblematic of his era in Europe and America. In Italy and France, in the late eighteenth and early nineteenth centuries, psychiatric patients in large public institutions were freed from the chains that bound them to their beds or to iron rings in the floors and walls. Seen more humanely than in the past, they were spared, to some degree, beatings and lashings by staff. In England, Quakers created a retreat where the patients gardened and sewed, played chess and dressed up for tea parties. And in

America, Benjamin Rush, physician, surgeon general of the Continental Army, signer of the Declaration of Independence, and close friend of Thomas Jefferson and John Adams, took charge of the psychiatric ward at Philadelphia's public hospital, where patients were kept in unheated basement cells and slept on straw on the floor. Rush persuaded the hospital to buy stoves for heat, then to build a completely new psychiatric wing. There was air; there was light; there were beds with proper mattresses. There were gardens to tend and grounds to stroll. An old Philadelphia tradition—Sunday outings to the hospital, to pay for admittance to the basement and the chance to gawk at and taunt the mad in their cages—came to an end.

Yet this humanistic trend was paired with a fierce drive to rectify—physiologically, materially, scientifically—the aberrations of the mind. The Enlightenment had lifted the idea that the melancholic and deranged were possessed by spirits. It had replaced the spiritual with the physical as the domain to be corrected or purged. A patient's system needed to be shocked; it needed to be arrested and restored; it needed to be rebalanced.

Benjamin Franklin—inspired by a Dutch scientist who wrote to him about an accident, a severe electrical jolt to the scalp, that left him feeling, the next morning, "the most lively joyce . . . a liveliness in my whole frame, which I never had observed before"—became a proponent of electroshock to treat melancholia. Rush, meanwhile, designed the Tranquilizer Chair, an unpadded piece of furniture, with an opening for waste, where the patient was tightly restrained for twenty-four-hour periods, with a wooden box over his head that was fixed to the chairback, keeping the torso rigidly erect, preventing head movement, reducing blood flow to the brain, and eliminating visual stimulation, all the while giving ready access for the massive bloodletting Rush swore by. The Tranquilizer provided a physiological reboot for the deranged.

Cures depended on a surrender by the patient, a reverence for the physician, the medical scientist. Treatments should both stun the system and induce utter submission. Rush's mentor, William Cullen, a renowned Scottish professor of medicine, described the necessary "awe and dread." An exemplary treatment, Cullen wrote, involved surprising patients by dropping them through a trapdoor and into an ice bath, then "detaining" them there, blending terror with "a refrigerant effect." Blistering was another way to break a patient down, while also draining defective fluids. Mustard powder was rubbed into a freshly shaved, raw scalp. When blisters formed, a caustic was massaged in. Patients reached up in agony and sometimes transferred the caustic to their genitals. The "indescribable" suffering, one doctor advised, helped the patient to "regain consciousness of his true self."

None of this was marginal medicine, let alone quackery. It was the prevailing science; it reflected the rationalism and optimism of the age. The imprint of the Enlightenment mixed the humanism of broken chains with an ardent, rigorous belief, a semi-secular faith, in the potential for human improvement. Mankind was capable of great progress. It could develop a government whose powers were carefully distributed, a structure that would serve the social contract and foster a stronger society and people. And it could devise a chair that would redistribute circulation and foster a healthier mind.

Beckoned to cure the king, Willis didn't employ the farming and fresh air that was part of the program at his asylum. The doctor told the court that his technique was to "break in" patients like "horses in a manège." He put the king in a straitjacket and bound him to a special chair. He set leeches on the king's temples to bleed him. Blisters were raised on the king's legs and became so badly infected he couldn't walk. Willis laced the king's food with toxins that made him vomit convulsively and plead for death. This went on for nearly three months.

It's impossible to say how often, in general, such methods

succeeded, or appeared to, but in the king's case, they caused, or coincided with, a remission. Suddenly he was himself again. The queen commissioned a "Prayer of Thanksgiving Upon the King's Recovery" that was read in churches throughout Britain, and Willis became a celebrity, painted by one of the most famous portrait artists of the day, rewarded by parliament with an annual payment of gratitude, and consulted by Rush, from across the Atlantic, about the physiology of madness.

The decades following Willis's triumph brought more means of gaining a patient's abject compliance and rectifying his system. Rotation therapy predated this period, but now patients were strapped into improved gyrators. They were spun at high speed for harrowing lengths of time. This induced obedience. And in the depressed, who were gyrated horizontally, with the feet at the axis point and the head making the widest revolutions, spinning was believed to direct much-needed blood to the brain. Unlike the deranged, who suffered from too much cranial blood, the melancholic had too little.

During this time, Joseph Guislain, a Belgian physician and psychiatrist—the term "psychiatry" having been newly coined by a German practitioner—created a method of stunning the patient and resetting the brain that relied on asphyxiation. Early in his career, Guislain won a competition held by the Belgian king to advance the treatment of the mentally ill. Soon afterward, in the late 1820s, he built what he called "The Chinese Temple." It was an iron cage, attached to ropes and pulleys, and contained within a mini facsimile of a Confucian house of worship. The whole structure sat above a small body of water. The patient was led to the temple and locked into the cage. Then, as Guislain described, a servant "releases a brake, which, by this maneuver, causes the patient to sink down, shut up in the cage, under the water."

)

PHINEAS GAGE WAS blasting a path for railroad tracks in Vermont, in 1848, when an explosion drove an iron rod through his head. He was a dependable crew foreman, and that afternoon he was doing what he always did. He set explosive powder into a hole that had been bored vertically into an outcrop. He ran a fuse to the surface and began tamping soil down into the hole, atop the powder, so the force of the blast would be channeled more outward into the rock than upward through the hole. But something distracted him; he turned toward his crew. His tamping iron ignited a spark against the rock. The premature explosion propelled the rod—an inch and a quarter in diameter and over three and a half feet long—into Gage's face below his left cheekbone, upward through his brain, and out the top of his head. It landed eighty feet away.

Gage lay on his back with his arms and legs in spasm. But within minutes, he was walking and talking and then riding in an oxcart, being driven to his hotel. A half hour after the accident, he was sitting on the front porch of the hotel when a physician arrived to examine him. He told his story to the doctor and onlookers before standing up and retching. "The effort of vomiting," according to the physician, "pressed out about half a teacupful of the brain, which fell upon the floor."

Though the tamping iron had rocketed right through the frontal region of Gage's brain, within a few months he was plowing on his parents' farm. He attempted to return to railroad work but instead began exhibiting himself for a fee as a roving curiosity around New England. He was examined at Harvard Medical School and presented to an elite society of Boston physicians. After his death, in 1860, his brain and tamping rod were donated to Harvard Medical School, where they are still on exhibit.

Gage's case is a landmark in brain science, not simply because he survived, but because the trajectory of his life suggested that discrete areas of the brain are primarily responsible for certain abilities and behaviors. Despite his injury, Gage's memory and most other faculties stayed intact, but his personality seems to have changed. A Vermont doctor who took over Gage's case at the outset reported, in what was later renamed the *New England Journal of Medicine*, that Gage's railroad employers, "who regarded him as the most efficient and capable foreman in their employ previous to his injury, considered the change in his mind so marked that they could not give him his place again. He is fitful, irreverent, indulging at times in the grossest profanity (which was not previously his custom), manifesting but little deference for his fellows . . . obstinate, capricious and vacillating, devising many plans of future operation, which are no sooner arranged than they are abandoned in turn for others appearing more feasible. In this regard, his mind was radically changed, so decidedly that his friends and acquaintances said he was 'no longer Gage.'"

The doctor's account has lately been challenged as overstating Gage's transformation, but the case is among the most influential in the history of neuroscience. It pointed to the prefrontal cortex—the frontmost area of the brain, running from directly behind the forehead to partway back beneath the skull—as the seat of what we now call executive function: the capacity to plan, check our impulses, process complex social information, filter our speech, make sound decisions, pursue high-level goals. The region has become known, too, as the locale of imagination and introspection, and been designated, for popular consumption, as the part of the brain that makes us human. The designation is loose, given all the far-ranging circuitry that is vital to the self and self-awareness, though it's true that our prefrontal region is a lot bigger than the same area in apes, and that neuroscientists debate whether mice have a prefrontal cortex at all. Gage and his tamping rod, at any rate, mark a foundational moment in the mapping of the brain.

In the 1860s and '70s, Paul Broca, a French surgeon, and Carl Wernicke, a German psychiatrist, performed autopsies on aphasiacs and, based on their brain lesions, isolated areas essential to language. From autopsies on patients who'd been afflicted with epileptic seizures, John Hughlings Jackson, an English physician, linked the movements of our bodies to what would be named the motor cortex. And working in a cramped hospital kitchen in the early 1870s, Camillo Golgi, an Italian doctor, developed a staining technique that revealed, under a microscope, in black against a yellow background, the axons and dendrites: the interconnected and interacting tails and tentacles running into and projecting from the bodies of the nerve cells—the neurons—of the brain.

Golgi won a Nobel. And in the late 1870s, as he wrote *The Brothers Karamazov*, Fyodor Dostoyevsky seems to have intuited the magnitude of what Golgi had done, that perhaps more than any earlier scientist, Golgi had rendered a vivid materialist vision of who we are. Without mentioning Golgi by name, Dostoyevsky has his character Mitya speak to his brother Alyosha—semi-articulately, in an anguished, frenzied outburst staggered with ellipses—about Golgi's breakthrough and how it leaves him feeling amazed and horrified and lost:

Mitya says, "Why am I lost? Hm! The fact is . . . on the whole . . . I'm sorry for God, that's why!"

"What do you mean, sorry for God?"

"Imagine: it's all there in the nerves, in the head, there are these nerves in the brain (devil take them!) . . . there are little sorts of tails, these nerves have little tails, well, and when they start quivering there . . . that is, you see, I look at something with my eyes, like this, and they start quivering, these little tails . . . and when they quiver, an image appears, not at once, but in a moment, it takes a second, and then a certain moment appears, as it were, that is, not a moment—devil take the moment—but an image, that is, an object or an event, well, devil take it—and that's why I see and then think . . . because of the little tails, and not at all because I have a soul or am some sort of image and likeness, that's all foolishness. Mikhail explained it to me, brother, just yesterday, and it was as if I got burnt. It's magnificent, Alyosha, this science! The new man will come, I quite understand that . . . And yet, I'm sorry for God!"

STILL, NONE OF these physiological advances, from Gage to Golgi, led to new psychiatric treatments. Problems of the psyche remained elusive in the brain. Autopsies of asylum patients could discern, in the lobes and folds, no physical markers of mental illness, no explanatory lesions, no telltale signs. Through the middle and much of the late nineteenth century, with gyration and asphyxiation abandoned as cruel, the main method of treatment was just to give the sufferer a rest, a break from society, respite that was viewed, at first, with abundant hope. Across the United States and Western Europe, laws were passed to ensure that more and more asylums were built. "It should never be forgotten," the director of a new pastoral refuge announced, that "every tree that buds, or every flower that blooms, may contribute in its small measure to excite a new train of thought, and perhaps be the first step towards bringing back to reason, the morbid wanders of the disordered mind."

But overcrowding, the construction of larger, less manageable institutions, and the tightening of public budgets meant that over the decades the goal of serenity gave way. One superintendent complained of having to fill his staff with "criminals and vagrants." The use of seclusion cells and restraints surged. Journalists feigned symptoms, got themselves committed, and emerged with exposés, like Nellie Bly's of a women's asylum, in 1887, on Roosevelt Island off Manhattan: patients beaten by staff with broom handles, choked, slammed headfirst into walls, straitjacketed, roped together, packed onto straight-backed wooden benches and made to sit silently for fourteen-hour stretches, given clean clothes once per month.

Throughout, psychiatry never quite lost its belief in somehow expunging the physical sources of the psyche's troubles—and in the very late nineteenth and early twentieth centuries, the effort intensified. Henry Cotton, the superintendent of a New Jersey asylum, set about curing patients by extracting their teeth, gallbladders, ovaries, testes, and intestines in order to remove the toxicities that were hiding in these locations and migrating upward, causing, he said, "mental darkness." He claimed an eighty percent success rate. He was championed by the field of psychiatry until an investigation discovered that his intestinal surgeries alone had killed over a hundred patients. Apparently inspired independently, without knowing about Cotton's approach, a Chicago surgeon tried to cure his son of psychosis by operating on his colon to free problematic fecal matter. The surgeon killed his son yet went on performing psychiatric operations. And soon it was comas that became the innovation of promise—comas, induced by excessive injections of insulin, from which patients were said to wake with their brains restored.

"I THINK BRAIN anatomy," Sigmund Freud wrote to his fiancée at the start of his career, in 1885, "is the only legitimate rival you have or will ever have." He hunched over microscopes, studying the nervous system of the sea lamprey, an eel-like fish, and a portion of the human brainstem. His life might easily have been devoted to the brain's material, but in the years just after writing that line to his beloved, his work turned in a different direction. He spent time on a fellowship, immersed in the condition then termed hysteria, whose symptoms spanned from dramatically altered personality to the loss of speech, and he began developing a method of excavating memories and uncovering unacknowledged desires. His clinical work led him to repudiate some of the leading brain anatomists of the era. In 1891, he published a critique of Wernicke, who had tied the comprehension of language to a particular spot in the brain and whose research was crucial to the reigning theory that specific human abilities could be attributed to precise cortical areas. Freud dismissed this as reductive. His instinct was that the brain operated in ways much too complicated for such straightforward mapping.

Over the next half century, until his death in 1939, Freud never denounced the potential of brain science. The problem, he said, was that the potential lay in the remote distance. "In view of the intimate connection between the things we distinguish as physical and mental," he wrote in 1926, "we may look forward to the day when paths of knowledge . . . will be opened up, leading from organic biology and chemistry to the field of neurotic phenomena. That day still seems a distant one, and for the present these illnesses are inaccessible to us from the direction of medicine." Despite the intimate connection, there was a "gulf," he continued, "between the physical and the mental."

Rather than commit his life to apprehending the brain, Freud dedicated himself to its intangible analogue, its inseparable yet distinct counterpart, the mind. He outlined an immaterial cartography. There were dominions of buried emotional experiences and unconscious psychological forces. The id, ego, and superego were the headquarters of our lustful and aggressive animal drives, our sense of self, and our socially imposed, internalized sense of morality, and the three coexisted within us in volatile, sometimes calamitous proximity.

The route to treating mental illness lay in listening, in revealing what was buried, denied, obscured. Dreams were direct or distorted flashes of fantasies. Defensiveness divulged yearning. Our untenable or taboo urges were redirected, if our psyches were reasonably healthy, into positive outlets through a process of sublimation. Though Freud and Freudianism have been maligned as absurdly unscientific and as misogynistic, and though there's good reason for both accusations (in one extreme example of both absurdity and sexism, Freud suggested that women naturally took to weaving because the intertwining of strands resembled their pubic hair, which in turn masked their missing penises), Freud's unprovable insights and constructions survive in the way we conceive of ourselves. "We remain Freudians in our daily lives," the psychiatrist Peter Kramer has written—though Kramer, author of the bestseller *Listening to Prozac,* may be America's best-known advocate for the biochemical comprehension and treatment of our psyches. "We discuss intimate concerns in Freud's language, using words like ego and defensiveness. We listen and observe as Freudians. As others address us, we make note of telltale incongruities that simultaneously hide and reveal unacceptable thoughts and feelings. When a lover forgets a rendezvous, we fear that otherwise unexpressed hostility may be at issue." Freud's layered psyche is full of "moral implications" about the need for "self-scrutiny" and the ultimate value of "authenticity," Kramer goes on, naming Freud "the inventor of the modern mind."

But even while they amassed influence in psychiatry, Freud's ideas didn't derail the materialists. In the mid-1930s, a pair of chimpanzees, Becky and Lucy, had their frontal lobes excised by two Yale neurologists, who presented their findings at an international conference. The chimps lost their ability to solve problems, but Becky, the more emotional of the pair, who had previously gone into tantrums when frustrated, was now so calm that she seemed, the scientists said, to have joined a "happiness cult." A Portuguese neurologist, Antonio Moniz, was in the audience, and within months he recruited a neurosurgeon to drill holes in the skulls of depressed, anxiety-ridden, or psychotic patients, and inject pure alcohol to destroy links between the prefrontal cortices and other areas of the patients' brains. Then Moniz and his surgeon improved on the procedure. This was the birth of the lobotomy: They dispensed with the injections and substituted a slender blade, which was inserted into the holes and rotated, coring the targeted brain matter.

Of his first twenty cases, Moniz reported that one-third were cured, one-third benefited, and one-third were unchanged. A Lisbon psychiatrist who supplied some of Moniz's patients said that the surgery was based on "cerebral mythology" and that it badly damaged personality, but criticism didn't stop lobotomy's spread in Europe, the United States, and Latin America. In the States, Walter Freeman, chair of George Washington University's neurology department, quickly became the operation's leading practitioner. His "surgery of the soul," the *New York Times* stated on its front page, in 1937, alleviated "tension, apprehension, anxiety, depression, insomnia, suicidal ideas, delusions, hallucinations, crying spells, melancholia, obsessions, panic states, disorientation, psychalgesia (pain of psychic origin), nervous indigestion, and hysterical paralysis."

"From problems to their families and nuisances to themselves, from ineffectives and unemployables," the *Saturday Evening Post* wrote, in 1941, patients "have been transformed into useful members of society. A world that once seemed an abode of misery, cruelty, and hate is now radiant with sunshine and kindness to them."

A few months after the *Post* story, Joseph Kennedy, wealthy investor, ambassador to the United Kingdom, and father of the future president, contacted Freeman. Rosemary, John's sister, was learning-disabled and prone to erratic moods. She was also beautiful and had a habit of wandering off from the convent school where she lived in her early twenties. Fearing that sexual scandal would ruin the family's political prospects, and without telling his wife, Joseph delivered Rosemary to Freeman. The lobotomy he performed left her walking with a limp and, cognitively, much worse off than she'd once been. Her parents hid her in a Wisconsin institution, where she stayed for the rest of her life.

Freeman was undeterred. He created a technique that eliminated skull drilling and made the surgery much easier and more available: pull back the patient's eyelid, insert a pick above the eyeball, tap with a hammer to drive the pick through the thin bone separating the eye socket from the cranial vault, wriggle the pick back and forth to get rid of frontal brain tissues, and repeat via the other eye socket. It was all done in less than five minutes. The patient could be home the next day. Freeman crisscrossed the country, executing and teaching his operation, and eventually lobotomizing patients in his van. Moniz won a Nobel. More than forty thousand lobotomies were carried out in the States, and ten thousand to fifteen thousand were done in Britain, before the procedure fell out of favor in the 1950s.

Freeman maintained that the great majority of his surgeries were successful but allowed that not all successful outcomes were perfect. After recovery, some patients were left, he wrote, at the "level of a domestic invalid or household pet." He noted that "every patient probably loses something by this operation, some spontaneity, some sparkle." But the trade-offs were worthwhile. The net result was tranquility. The surgery was "a stroke at the fundamental aspect of the personality," which was "responsible for much of the misery that afflicts man. . . . Even if a patient is no longer able to paint pictures, write poetry, or compose music, he is, on the other hand, no longer ashamed to fetch and carry, to wait on tables or make beds or empty cans."

TO POINT AT the past in horror can be a way to avoid pointing at ourselves. This was what Blessing, a psychiatrist and neurologist from Australia and Goff's young colleague at NYU, seemed to be saying when she mentioned the era of lobotomy and added that barbarity remained part of the standard treatment for psychotic patients. She wasn't equating then and now. She wanted that to be clear. We no longer went in with a pick and hammer. But the medications psychiatrists routinely prescribed—not just routinely but almost invariably, and not just prescribed but pushed as necessary or forced on the patient if involuntary treatment was a legal option—frequently sent the metabolism into disarray, made blood pressure soar, and caused the body to bloat or balloon. The drugs partially reduced psychotic symptoms in some, yet they also upped the odds of heart disease and diabetes.

The drugs dimmed thinking and dulled speech. They obliterated sex drive, stirred excessive salivation and drooling, and caused tardive dyskinesia: unrelenting, repetitive scrunching of the eyes, puffing of the cheeks, twitching and twisting and smacking of the lips, darting, lolling, and rhythmic thrusting of the tongue, jerking of the neck, jerking and gyrating of the shoulders, wringing of the hands, rocking back and forth. There was, too, in addition to these unwilled movements, the nearly uncontrollable pacing impelled by akathisia, a feeling of overall torturous physical restlessness that begged perpetually for release.

As I sat with Blessing in her small office, down the hall from Goff's, I said that I sometimes worried about being sensationalist about side effects. But then again, as we both knew, the percentages were significant. With Risperdal, one of the antipsychotics relatively unlikely to cause tardive dyskinesia, you still had a ten to fifteen percent chance of winding up with the condition. It tended to be irreversible. So the patient, who was typically diagnosed and medicated in her late teens or twenties, might be condemned to a life with her face or body besieged by tics and spasms, or both her face and body overtaken, a life of feeling monstrous, of enduring the staring eyes of children and the averted eyes of adults, all of this compounding the isolation and fear that came with the psychiatric disorder itself. And tardive dyskinesia was just one of the risks that loomed; the ten to fifteen percent didn't account for the other hazards. Risperdal came with the peril, for boys and young men, of gynecomastia—of growing unmistakable, full, even pendulous breasts. Johnson & Johnson, whose subsidiary made Risperdal, had paid three billion dollars to settle lawsuits about the medication. This was an acceptable cost of doing business, it seemed, given the tens of billions in sales Risperdal had brought in.

"Would you let your loved one take that drug?" Blessing asked.

I replied that the vast majority of psychiatrists would say that my loved one *should* take it—or another drug like it.

Blessing's office was faintly off-kilter. A lampshade was slightly askew, and she had warned me against sitting in a "trick chair." She herself was impeccably put together in a funky black ensemble, but in this, too, she was at odds; her stylishness made her stand out among the neuroscientists I'd met. She'd said that her husband was the founder of the Moth, the quirky storytelling cabaret. She wasn't quite coming to her work with a conventional perspective.

"We don't have the ethical framework we need," she said, meaning that psychiatrists didn't know how to weigh the factors that should go into decisions about medication. She detoured onto the topic of temperature and a drug she envisioned. She spoke about a phenomenon known as brown adipose tissue. This type of fat—distinct from the whitish variety—has a special purpose, and babies have a lot of it. Brown fat cells can be easily burned and converted into heat. Babies need this protection against hypothermia more than older humans do, or, anyway, the babies of our ancestors did: babies' proportionally larger heads are exceedingly vulnerable to heat loss, and they can't readily move themselves from colder to warmer areas.

As we grow up, our supply of brown fat dissipates gradually. Meanwhile, when given antipsychotics, children and young adolescents are even more prone than adults to major weight gain. Blessing talked about how there might be a connection. But she veered, first, onto the topic of pharmaceutical companies pushing psychiatrists to prescribe—and using public awareness campaigns to get parents to request—antipsychotics such as Risperdal and

Zyprexa for kids, not for psychosis but for behavioral problems, because of the drugs' sedating effect. She then said there was fledgling evidence that antipsychotics dial down the brown adipose heating system, leaving the young, especially, with excesses of this particular fat, which might be part of why kids on the drugs were so susceptible to obesity.

Blessing's thoughts about temperature had led her to a theory about medicating psychosis without the devastating side effects. There was evidence that psychotic flare-ups are accompanied by increases in brain temperature. And Blessing had done research suggesting that antipsychotics cool the brain down. These temperature patterns were tied, she said, to a set of tightly interwoven areas in the brain involved not only in the regulation of heat in the head and body, but also in the determination of what, in our environment, does and doesn't warrant our attention. These intertwined areas—labeled the salience network—contain our metaphoric thermostat and are essential, along with the hippocampus, in assigning levels of significance to all we perceive.

I asked if there was an evolutionary explanation for why these two functions are so closely aligned in our brains. "Oh, absolutely," she said. Self-protectively, when animals recognize a threat in the environment, they draw blood and heat from their extremities toward organs that are vital to survival. We humans are programmed to do the same. And this happens in acute psychotic episodes. When she'd worked in hospitals and clinics, she'd noticed it in patients in crisis. Their fingertips were frigid. For them, subjectively, dire salience was everywhere; they were in emotional and physical danger, surrounded by hostility, infiltrated by voices of condemnation, unable to filter out or fight off what they were hearing and seeing.

Blessing's hunch was that in psychosis, the rise in brain temperature was more than just a *result* of misperception; instead, it might be part of a loop. It might both stem from and feed mistaken

salience. Her suspicion was that the downtick in brain temperature associated with the antipsychotics was relevant to reducing
psychotic symptoms rather than being a trivial side effect. She
wondered about developing a drug that targeted brain temperature specifically, instead of affecting it while crudely assaulting the
brain in all sorts of ways, as today's antipsychotics do. Could such
a drug play a role in treatment? Might such a drug be far less
damaging than the current barbaric medications?

AT TIMES, WHEN I listened to Blessing, or to Goff or Nestler, my
head felt clotted with all I struggled to understand, with infinitesimal subregions of the brain that were new to me, with the myriad
known and conjectural interconnections between these areas, with
the surmised purposes of these ever-altering webs made of the
hundred billion neurons that compose the material strata of our
minds. But I think I had their empathy. They, too, were at a loss.
Years and years of knowledge ahead of me, full of passion—and,
still, expectation—but at a loss. Goff and Nestler had begun their
careers at a time when comprehension had seemed to be waiting
for them, when solutions were lying somewhere close in their labs,
waiting to be seized. Now, as I sat with Goff in his spacious, orderly office up the hall from Blessing, his blond wooden desk and
round table burnished and immaculately bare, he thought about
one of his recent experiments, gave me one of his rueful laughs,
and said, "I understand less today than I did last year."

Nestler's office was another zone of order, as if to counter the
persistent blurriness, the unparsed mysteries, of his life's work. He
remembered being in medical school, forty-odd years ago, and
professors predicting imminent, precisely tailored cures for cancer. "And cancer is so dumbass simple compared with disorders
of the brain. We were so naïve."

Above his COVID mask, his eyes had a quality I saw, too, in

Goff's, a mix of sharp focus and keen humility. He spoke about the multiple sources of depression and anxiety, then elaborated on one of these sources, the genetic, and recalled an early, false reckoning he'd had with complexity: "We thought, okay, depression is genetically complicated, so instead of one gene there will be five. Well, there are probably a thousand, each having a tiny effect."

Yet for all his frustrations, he was quick to invite me to visit his labs. He was eager for me to take in what he was discovering about resilience, about the capacity of some to fend off depression, fend off debilitating anxiety, eager for me to grasp his lessons about the interaction of the meta with the minute, the environment and upbringing with the molecular machinery that is constantly creating and adjusting connections in our brains. He wanted me to observe his laboratory mice and consider the persistent and often lifelong differences—the mechanistic differences in the brain—between mice who were sufficiently licked and adequately nested in childhood and those who weren't so lucky.

He'd had a eureka moment in the mid-2000s. It happened after listening to a colleague present his research on prisoners of war, his interviews with Americans who'd been POWs in Vietnam, like John McCain, men who'd endured years of torture and done well afterward. Would it be possible to illuminate the neuroscience of resilience? To find the physiological basis of this capacity? To translate this somehow into a medication? This was one of the projects Nestler and his team were engaged in now.

BOTH NESTLER AND Goff were products of a battle that culminated around 1980. On one side of the battle were two unallied factions, the descendants of Freud and a movement of relativists—factions that understood the psyche in immaterial terms. On the other side was an alliance of biologically minded psychiatrists and the giants of the pharmaceutical industry. The psychiatric treatments that my brother and David and Caroline received were shaped by this battle as well, and the same is true, today, about the emphasis on psychotropics that we continue to take for granted.

Freud, his composition of the psyche, and his system of talk therapy, psychoanalysis, didn't gain their full influence until after his death. In America, at the outset of World War II, the military put a Freudian in charge of weeding out draftees and volunteers who might be mentally unfit to serve, a process that included ferreting out homosexuals. After the United States entered the war, a psychoanalyst was named director of psychiatry for the Army. He was responsible for designing programs that would protect soldiers from breakdowns at the front. Men who did snap—men who suffered from what had been termed shellshock in World War I but was now diagnosed, borrowing Freudian language, as war neurosis—were placed under the care of a former student of Freud's. The broken were treated with a kind of high-speed psychoanalysis. It started with an injection of a trance-inducing drug that was considered a shortcut to the unconscious and to disinterring the troubles that lay there.

With the war over, President Truman sent a message of greeting to the annual conference of the American Psychiatric Association, with its leadership of Freudians and neo-Freudians. He hailed the group as integral to the nation's—and the world's—most urgent wish: "The greatest prerequisite for peace, which is uppermost in the minds and hearts of all of us, must be sanity." By the early 1950s, psychoanalysts chaired most of the psychiatry departments across the country, and toward the end of the '60s, the autobiographical novel *I Never Promised You a Rose Garden* was adopted into high school and college curriculums. The book, which would sell millions of copies, romanticized the analyst, Frieda Fromm-Reichmann, who had liberated the author, when she was a teenage girl, from the grip of alternate realities and suicidal drives.

The relativists had a different impact. Three of the most consequential thinkers of the 1960s and '70s, the philosopher Michel Foucault and the psychiatrists R. D. Laing and Thomas Szasz, attacked the very foundations of the profession. They argued that the diagnosis of mental illness was based on nothing objective, that it depended completely on culturally determined definitions, boundaries, norms. Foucault taught that psychiatry's claims to rationality and specialized knowledge were inherently deceitful, that they were a means to stigmatize, quarantine, and subdue those who think with originality, those whose visions might disrupt the prevailing social order, those capable of "joining the figures of night to the powers of day, the forms of fantasy to the activity of the waking mind." Laing saw psychiatric classifications as soul-killing: "I would wish to emphasize that our 'normal' 'adjusted' state is too often the abdication of ecstasy, the betrayal of our true potentialities, that many of us are only too successful in acquiring a false self to adapt to false realities."

Szasz, in *The Myth of Mental Illness*, shone a light on a void: the absence of anything physical, anything identifiable in the human brain, beneath psychiatry's assertions of illness. His allegations of the profession's emptiness became so popular that they were reprinted on posters and adorned dorm room walls. "Classifying thoughts, feelings, and behaviors as diseases," he declared, "is a logical and semantic error, like classifying the whale as a fish." He equated psychiatry's oppressive effects with animal predation: "In the animal kingdom, the rule is, eat or be eaten; in the human kingdom, define or be defined."

Feminists leveled similar charges. There was nothing objective, Betty Friedan wrote, in psychiatry's tendency to blame pathologies on poor mothering. And there was something blind—or worse—in the profession's failure to tie female depression to the submissiveness inflicted on women by society and imposed as a norm by psychiatry itself. And why, homosexuals demanded to know, following the 1969 Stonewall riots that ignited the fight for gay rights, was same-sex attraction categorized as a mental illness? What exactly was the rationale? Wearing a Halloween mask, a wig, and a floppy tuxedo, and using a voice distorter, an anonymous member of the American Psychiatric Association spoke at the group's 1972 convention, stating, "I am a homosexual. I am a psychiatrist," and compelling the profession to confront the irrationality—and harm—of its diagnostic criteria. Psychiatry's cruelty and repressive paradigms were indelible themes in the movie version of *One Flew over the Cuckoo's Nest*, which swept the 1976 Oscars, winning for best picture, director, actor, actress, supporting actor, cinematography, editing, adapted screenplay, and score.

The profession felt itself under existential threat. For American medical students picking a specialty in the years after World War II, psychiatry had been the fastest growing discipline. During the '70s, the percentage of new doctors choosing psychiatry dropped by more than half. At the same time, the practice of talk therapy was being overtaken by psychologists and social workers, by lesser-paid and lower-status professionals without medical degrees. Psychiatry needed to reassert its superior knowledge, its unique authority regarding the mind, its objectivity, its place as a true science.

The profession's biological wing published a manifesto in 1978: "Psychiatry is a branch of medicine. . . . There is a boundary between the normal and the sick. . . . Mental illnesses are not myths. . . . The focus of psychiatric physicians should be particularly on the biological aspects of mental illness." When the American Psychiatric Association released, in 1980, a third edition of the *Diagnostic and Statistical Manual of Mental Disorders*—the profession's bible of illnesses, its reference guide for categorizing patients and billing insurance companies—a claim of hard science was made on every page. It was there on page one, with the pledge of "an increased commitment in our field to reliance on data as the basis for understanding." It was there in the sheer number of pages. The previous editions, the DSM-I and DSM-II, put out in 1952 and 1968, had each been around one hundred and thirty pages; they'd barely qualified as books. The DSM-III was a weighty volume approaching five hundred pages, every page newly dense with text.

The DSM-III rejected the immaterialism of Freud; it dispensed with Freudian language and concepts. Unconscious turmoil was repudiated as a source of symptoms. The book's framers couldn't point to proven neurological causes or material diagnostic markers for psychological illnesses, but instead, they lent the DSM-III the aura of empirical science by adding to the types of diagnoses that could be given—there were three times more than in the DSM-I—and by filling the book with extensive checklists, like this one for a type of depression:

During the depressive periods at least three of the following symptoms are present:
1. insomnia or hypersomnia
2. low energy or chronic tiredness
3. feelings of inadequacy, loss of self-esteem, or self-deprecation
4. decreased effectiveness or productivity at school, work, or home
5. decreased attention, concentration, or ability to think clearly
6. social withdrawal
7. loss of interest in or enjoyment of pleasurable activities
8. irritability or excessive anger (in children expressed toward parents or caretakers)
9. inability to respond with apparent pleasure to praise or rewards

10. less active or talkative than usual, or feels slowed down or restless
11. pessimistic attitude toward the future, brooding about past events, or feeling sorry for self
12. tearfulness or crying
13. recurrent thoughts of death or suicide

Looking back, it's easy to wonder if this new diagnostic approach, which was both systematic and highly elastic—how many people wouldn't meet at least three of these criteria at times in their lives?—was created in collusion with the pharmaceutical companies. There's no evidence of outright conspiracy between the academic psychiatrists who composed the volume and the corporations that were, by then, manufacturing plenty of drugs, and developing plenty more, to be prescribed for the diagnosed. Yet a courting of psychiatrists by the industry was already well under way. And there was certainly a convergence of agendas. Pharmaceutical remedies—molecular solutions—meant spectacular profits for the companies, and, for the profession, coveted authority, seeming objectivity, a mantle of pure science.

WHENEVER I SPENT time with Goff or Nestler, I felt torn. It was thrilling to enter their minds, which were, in turn, traveling to greater and greater depths in our brains. It was thrilling to hear Goff and a collaborator, Lila Davachi, a neuroscientist at Columbia University, describe how our memories, along with our constructions of reality, might take shape and take lasting hold along the webs between a segment of the hippocampus and a portion of the prefrontal cortex, and to hear them delineate how the alternate realities of psychosis might form and flourish—and become difficult to dislodge—along these pathways. And it was thrilling to have Nestler guide me through the internal signaling within the cells that compose the networks. Here was the substantive foundation of our subjective perceptions of ourselves and all that surrounds us.

To be in their presence was to be immersed in the mind-brain questions that have entranced thinkers since René Descartes, who sketched elegant vectors in his attempt to capture, in the seventeenth century, the relationship between the physical realm inside the skull and the intangible soul. To be with Goff and Nestler was to delve, privately as I listened, into questions that divide today's philosophers and scientists of artificial intelligence; questions that separate anyone with even vague flickerings of spirituality from those who keep the ineffable at bay; questions that, if we allow ourselves to dwell on them, won't leave us alone. How does the physicality of light, and our merely mechanical processing of it, become a painting of singular and mesmerizing subjectivity by Turner? How do grass, trees, and water become these aching yet transcendent lines, about time and aging, by Wordsworth:

> *There was a time when meadow, grove, and stream,*
> *The earth, and every common sight,*
> *To me did seem*
> *Appareled in celestial light,*
> *The glory and the freshness of a dream.*
> *It is not now as it hath been of yore;—*
> *Turn wheresoe'er I may,*
> *By night or day.*
> *The things which I have seen I now can see no more.*

How does the juxtaposition of tones, which can easily be charted mathematically, become the compositions of Mozart or the singing of Billie Holiday, music that is, for some, a suggestion, a frail hint or an almost undeniable manifestation, of God?

But to be with them made me uneasy. This was not because

they had worked closely, in the past, with the big pharmaceutical corporations, consulting for them, pitching them ideas for new
medications, administering their drug trials, nor because Nestler
still had a fair bit of involvement with smaller-sized companies and
had founded a company of his own. Their candor and self-criticism
spoke to their ethics. And, importantly for me, they were engaged
by the philosophical questions that bordered their work; they were
given to contemplating the dilemma of how circuitry becomes sen-
sibility and how cells become personhood, to considering the leap
between levels of being. Yet they were far less compelled by this sort
of thinking than by the microscopic machinery.

My uneasiness vibrated most with Goff. This was because, despite his humility, and despite his pursuit of better and safer medical treatments, he was confident about the wisdom of prescribing
some of the most destructive drugs we have, prescribing them as
the default method, confident that, on balance, intervening in the
brain with these pills or injections is the best course in situations
of psychosis, and that the intervention should be made early, at
the first clear signs of disorder, often when the patient is still in
his teens, before anyone can tell how her symptoms will play out,
whether they will intensify or fade over time. He was confident
that the risk of the illness getting worse should be addressed with
pharmaceuticals, never mind the element of barbarism.

Sometimes I was too thoroughly enmeshed in his thoughts, in
the pathways he was mapping, in his combination of humility and
expertise, to worry. But sometimes I imagined a scene that was
not far-fetched: my brother being evaluated by him, behind the
heavy, locked doors of a psych ward in 1983. Goff with his striking
blue eyes. Goff in a dark blue suit. To my brother, he would have
looked not as he seemed to me: reflective, self-critical. He would
have looked, with those doors bolted, omnipotent, ominous. And
Goff would have seen my brother just as my brother was, in fact,
seen, as someone whose brain was badly diseased.

THREE

Families are subjective spaces. It is true that the same idea can be applied to anything, that we experience everyone we interact with and everywhere we inhabit in our own ways, but within families our divergent perceptions are most unsettling; we see the same people up close, feel them so powerfully, yet so differently. The divergence can be too difficult to accept. Even if we allow ourselves to examine it, we have to turn away from it, to preserve our own vision, our own knowledge. Especially with our parents, easier to have a fragmentary and flawed understanding than a complete and contradictory one. Better to tell ourselves our own stories. Easier that our parents remain well-outlined and well-defined, almost like fairy-tale figures. Better not to ask too often, Who *was* my father? Who *was* my mother?

When we moved from Brooklyn, when I was twelve and my brother was ten and a half, we chose a house at an outermost edge of Seattle, in a small neighborhood where the city jutted bluntly into Puget Sound. It was the early seventies. When I had told my Brooklyn friends we were going to Seattle, their only association was with a TV series called *Here Come the Brides*, about marriageable women shipped to the semi-wilderness of Seattle in the nineteenth century. We moved when Starbucks was a single store and Microsoft was unconceived. The members of what would be Nirvana were in elementary school or kindergarten. Seattle was Boeing, the airplane builder, which had just laid off three-quarters of its eighty thousand workers. Seattle was Weyerhaeuser, though the timber company's headquarters was actually in Tacoma, thirty-five miles away. Seattle was a downtown where one of our father's worries, as the new director of the health department, was how to help the legion of hardened, grizzled alcoholics, among them lots of Muckleshoot and Duwamish, who staggered in the winds off Eliot Bay.

The city had pockets of charm, and it possessed some of the self-congratulation that soon proliferated, but there was desolation in the air. And it was a place where a family of Brooklynites, particularly a family of liberal-minded Jews, could feel like aliens. The house we chose, in that neighborhood of thirty or forty homes, compounded the feeling of being cut off.

There were reasons for our choice. The half-acre of land around the house was magical, and, because it was not in one of the city's fancy sections, the place was affordable. When the four of us first visited, I couldn't believe it was in our price range. The property, near the crest of a steep hill, had four tiers of lawn divided by evergreens and tall shrubbery that were seemingly four separate worlds, the lowest one, farthest from the house, feeling on the verge of wild. There were various little gardens that the seller had kept carefully tended, and tucked into a nook was a pond, twelve feet across, spanned by a bridge of dark stone, with plump fish, blue and gold, drifting back and forth beneath the bridge.

The house dated back to the thirties; we all liked that it had some history. Wooden beams crossed the vaulted living room ceiling. French doors opened onto a front patio and a back porch.

Each bedroom was big enough to encompass two of the bedrooms in our apartment back in Brooklyn, and one of the windows in the master bedroom looked out not only on cedars and firs but on a sliver of the Sound. I was smitten. Our parents were as well. Bob was more quiet. He recognized right away just how isolated this place was. In Brooklyn, down the block, there had been stoop ball and street hockey and football with steam vent grills to mark the end zones. We had only to look out a window to know if friends were gathering. In this Seattle neighborhood, there were no signs of other kids. Our father asked the owner, who said he didn't think there were any families with children. In retrospect, it's easy to see that it wasn't an area that would attract people with any interest in community. Each house made a wordless announcement of privacy, and the neighborhood was bounded by the tall, sheer bluffs that stood above the sound and by a mostly defunct military fort.

At school, my brother and I were among the smallest boys in our classes. This was mostly because we each found ourselves skipped a grade, into seventh, into eighth, placements not necessitated by any natural gifts but rather due to the fact that we had attended a private school in Brooklyn and had covered more academic ground than the kids in our public junior high. In addition, my brother's late birthday had made him among the youngest even in his original grade; he was younger by almost two years now. But there seemed to be another factor at work; our shrimpiness wasn't only the product of age. Boys just seemed to be made bigger here—taller, stronger, beefier—possibly because, while our Brooklyn school, though secular, was filled with Ashkenazi Jews, our new school was filled with the descendants of hearty Germans and Norwegians.

There was a further issue. Our Brooklyn school had been progressive, cossetted, precious, populated by families whose dinner table discussions, in those Vietnam years, were about pacifism and conscientious objection. Here the boys challenged one another

to fights. Promptly at the three o'clock bell, the bouts were conducted near the school's parking lot or by the backstop of a dusty
baseball diamond. There was always a circle of cheering students.
The fights went on until one of the boys surrendered, saying, at
last, two scripted and repeated words: "I give, I give." We'd never
encountered anything like this.

At the price of some dignity, we managed to avoid fighting.
But my brother paid a higher price. Slight though I was, I was
somewhat adept at throwing a football and shooting a basketball,
and this gave me a foothold in the life of our new school, while
Bob, the more gifted all-around athlete, wasn't as skilled at these
mainstream sports. And while I could be boisterous, he was timid.
So, scarcely more than elfin in size compared with his classmates,
he disappeared.

To be "jewed" was a commonly used verb at our school.
"Jewed down" was a frequently used phrase, as when someone
had paid too much for a poster at the Pike Place Market or been
underpaid for a lawn mowing job. We heard this sort of thing a
lot. I don't know how cognizant the other kids were that we were
Jewish, how much they meant to taunt us by saying things, to
us or within our earshot, like "That old geezer tried to jew me
down"—or whether they meant to taunt at all. It's possible that
they weren't aware, or knew and forgot, because there were so few
Jews in Seattle then, and because this was just the way they and
their parents talked.

We didn't protest. Early on, Bob, confused, tested out the verb
at home on our mother, thinking that maybe it wasn't a problem,
that maybe it was just a word and he didn't have to feel uneasy
about its usage around him. He slipped it into a sentence. She
looked stricken.

"What did you say?"

"I said, 'He chewed me out of it.'"

She reprimanded him, gave him a concise lesson on anti-

Semitism. For her, it must have been another stark reminder of how far we'd moved from the city she knew and from the borough where she'd grown up, the borough her mother, Grandma Clara, had settled in after arriving in America as a young woman in the years between the World Wars, having left most of her family behind, spread between a town and a set of mountain hamlets near what is now the nexus of Ukraine, Romania, and Hungary.

Yet our parents didn't seek out the scattered Jews of Seattle. They didn't join a temple, let alone sign us up for bar mitzvah classes. My brother and I had no cluster of Jews to counter the chorus of "jewed" and "jewed me down." Our parents did not provide us with a Jewish identity; in a way, they were in flight from their own. In Brooklyn, our nuclear family had stayed away from large gatherings of relatives; my brother and I had never set foot inside a synagogue. Our parents were people of science. They had met at Brooklyn College, then ascended toward advanced degrees at NYU and Columbia and Berkeley. In their eyes, their extended families, especially on our mother's side, were people of superstition and inferior education, people who, whether or not they were strictly observant, believed in the irrational and were still tainted, thwarted, by their Hasidic origins in the Old Country.

This makes our parents sound like snobs. But there were forces, aftermaths, from which they felt they needed to wrest themselves free: periods of childhood poverty; a vanished father; insular upbringings; and—hardly ever mentioned in our house— the Holocaust.

In an enclave of immigrants not far from Coney Island, Grandma Clara was married and raising our mother while Clara's mother and five of her eleven brothers and sisters were gassed or otherwise slaughtered as part of Hitler's Final Solution. The cousin who first taught my brother and me to play catch—his mother and father had survived the camps. His mother, Shaindy, a warm, pillowy woman we kids liked to be around, had yearned to electrocute herself against a perimeter fence, rather than endure another day in Auschwitz-Birkenau. For our family, there was nothing abstract or remote about Holocaust history, yet my brother and I learned no details whatsoever as we grew up. I doubt we could have named the number of Grandma Clara's siblings or the total number of Jews who were exterminated. Shaindy's desire to kill herself was something she told me, along with hundreds of other facts, much later. At home, Bob and I were given nothing beyond a vague sentence or two. The most shattering event of the twentieth century might have happened at twenty degrees of remove.

This wasn't because our parents were trying to protect us. We had conversations about other painful subjects. They were trying, I suspect, to protect themselves. Their entire beings were invested in, wrapped up in, a faith in the rational, yet here was this behemoth that defied reason. That the Final Solution was viewed, by some historians, as an extreme and horrendous expression of reason itself didn't help. The Holocaust—this unequaled monster that had devoured our mother's family, that was packing her grandmother, aunts, uncles, and cousins into cattle cars when she was around eleven and doing her homework at Clara's kitchen table—defied explanation. Explanations had been attempted and fallen feebly short. The event could seem to demand not a rational dissection but a religious worldview and a religious term with religious capitalization, Evil, and so thoughts about the magnitude and intimate impact of the Holocaust must have felt overwhelming to our mother, to both our parents, something to be allowed into the mind for fleeting seconds before being pushed away and barred.

Our father, meanwhile, didn't like to dwell on anything outsized, whether as gargantuan as the Holocaust or as potentially uncontrollable and destructive as longing, as memory. It was impossible to get him to reveal much about his past. He shunned our questions, rebuffed us with a show of disinterest, as if there were nothing to tell. But he had a story of his own from the war years, a story that our mother had told me.

At the same time that our mother's family was packed into cattle cars, our father's father left his wife and only child, left without a word, and wasn't heard from for over a decade. This occurred on the eve of our father's bar mitzvah. Our father, thirteen, a diligent student, was making final preparations for his speech to the congregation. He had already picked up signals about the precariousness of his parents' marriage, and he may have believed that his performance, his biblical interpretation, if perfect, could make everyone happy and set things right. He knew that his mother was known, among her relatives, as the Ugly Duckling. She was one of six sisters, just like in the Hans Christian Andersen fable. And she was the homely one. It was considered a minor miracle that she was married off.

But his preparations accomplished nothing. His father didn't appear in the synagogue on his bar mitzvah morning, and didn't appear at home that night or ever again, and soon it was established that for some years he'd had another life, with another wife and other children in another state. Our father's parents' marriage had been too good to be true. This was the lesson our father absorbed. His mother was too unattractive to be loved. It was a lesson with the chaos of lust underneath it, a whisper teaching that desire for an alluring woman had led his father to abandon him and crush his mother, that needs beyond rationality lay always in ambush.

Our father lived in reaction to this. He was, in my eyes, a weak and painfully vulnerable man. How I hate writing that word—"weak"—even now. How I want to retract it, replace it by rushing to say again that in his work he emanated a reserved authority. How I wish to capture the precise sound of his level voice on the radio, or the steadiness of his large brown eyes on the local

TV news, as he explained to the people of Seattle why they should take this or that precaution to protect their health. How I'd like to describe, in great detail, a trip he took to Washington, D.C., with me along for a reason I can't recall. The two of us were welcomed into the inner, leathery, lamplit chamber of a U.S. senator, where our father and the senator discussed the urgency of a piece of public health legislation.

But weakness and vulnerability were what I felt. Anxiety crouched within his deliberate presence. He was struggling perpetually to keep anarchy quelled, to put his father's disappearance behind him. He had chipper sayings that he sprayed around the house. He invoked one of these tidbits whenever my brother or I didn't want to wear some ungainly item, an unfashionable parka or poncho, say, that would keep us warm or dry. "Life isn't a beauty contest," he would tell us. We heard this countless times, as if he thought that repetition could turn the words into wisdom. It had the opposite effect; I detected something suspicious. *Wasn't* life, in lots of ways, a beauty contest? Didn't the superficial assert itself an awful lot?

"Slow and steady wins the race" was another of his mantras, as was "Everything in moderation." These, too, raised questions in my mind, which would have been fine, had I been able to debate him. But I couldn't, not because I didn't have the wherewithal to make my points at the age of twelve or thirteen, but because I was afraid he would wilt, that he was incapable of defending his maxims, that my doubts, if expressed, would destroy him, that he needed to believe in his aphorisms of rationality and control, and needed us to believe in them. He loved the upbeat numbers from old musicals. Over and over he sang, "You've got to accentuate the positive. . . . Eliminate the negative." Each bouncy note contained more effort than exuberance. Each note confirmed his fragility.

Yet my brother had a different father, strong and looming. I can't make sense of it. He should have felt a deep affinity. Their distraught tears after we returned Ursus, their common empathy for that rotund, ebullient, hard-to-handle German shepherd puppy whom we'd failed, their common disappointment in themselves—this was just one sign of all they shared. Together after that, they watched over the bird feeders outside our kitchen and below one of Bob's bedroom windows, making sure that the robins and warblers, sparrows and nuthatches had enough to eat.

They both played the piano. They took lessons from the same jazz pianist. They practiced on the same upright, adjacent to the same French doors. Our father practiced a springy big band tune, "Hot Toddy," with catchy, cheerful right-hand runs and trills. Bob played a heavy, thrashing blues, with a pounding, unremitting left hand and a right that swept higher and higher, and struck harder and harder, driving at the upper half of the keyboard in an improvised rendering of solitude and release, before his fingers sped up into a hurtling of ecstasy. This was during a period when he had turned away from classical, which he'd studied from a young age in Brooklyn. I doubt he weighed a hundred and five pounds, but he made the instrument tremble and the house feel nothing less than violently occupied. He stoked himself into a blues-based trance.

There's plenty to be said about their opposite musical styles, beginning with the fact that our father absolutely did not improvise, didn't so much as attempt it. He adhered to the notes on the page, never veering, never adding so much as a momentary flourish in the middle or a repeated, fading measure at the end, his obedience meaning that whatever song he was working on was, from start to finish, quite short, and that he replayed it many, many times in a session of practice. Or, for more contrast between them: our father's light touch on the keys. Still, they shared a love of the instrument, a teacher, and jazz-infused repertoires. And later, when Bob added dance to his ambitions, he began in a genre, tap, that was a passion of our father's, that our father loved to see performed and may even have performed a bit himself as a teenager, when he had a summer job as a singing waiter in the Catskills. It was as if my brother was drawn to the piano and tap because they were ways of reaching out to our father, impressing him. On some deep level, their sensibilities were tightly tied.

At night, though, at bedtime: a confluence of architecture, animals, psychology. My brother's room had been built onto the main house; it projected outward toward the trees. It was exposed on three sides. He was alone. In the dark, raccoons took over the roof above his bed, darting in frenetic paths, claws scratching. Their throats gave out haunting screeches, hisses, growls, whimpers. They were strange communications, menacing. He knew their source yet could never be sure. Then our father approached along the short hallway between the original house and my brother's annex. The light was behind him. His long shadow extended ahead of him. It entered the room before him. So, in my brother's sight, below the close canopy of hostile sounds, there was the shadow and the body only afterward.

This, not the fragility I perceived, defined our father for my brother. He towered, with a reach beyond his reach, with an almost supernatural aspect to his power. Can both fathers, the delicate and the imposing, have been real?

A parallel question could be asked about our mother, who taught me that I could be anything I chose while imbuing my brother with the feeling that his dreams were delusions. I'm thinking ahead a few years to high school and beyond. Among her colleagues, at the dinner parties she threw, she proclaimed that I was going to be a novelist. Insisting on everyone's attention, she advertised my teenaged accomplishments and put me on display. She made no grandiose predictions and bragged with no giddy abandon about my brother. My future was limitless and artistic; his was uncertain and most likely mundane.

This dichotomy—the excess of support for me, the doubt directed toward him—was in the air before high school. Her appraisals were intimated, imparted. I did my best to maintain the imbalance, dominating our dinner table conversations with attempts at intellectualism that were meant to, and did, draw her into fervent debates. To think back on those absurd arguments is to cringe, but they had the desired effect. My brother, who always sat to my left, while our mother and I sat across from each other, might as well have been relegated to another room. He was mute; he was expunged; and our mother seemed oblivious to what I well knew was unfair and felt as assaultive, though I had no inclination to stop myself and would have been desolated had she cut off our seemingly high-minded discussions and turned her eyes to him.

She didn't. She was not one for moderation or balance. She left self-restraint to our father. She never tried to equalize her gaze or affirmations. Her capacity for self-control may have been too depleted by the effort it took to deal with an illness that had surfaced when she was sixteen.

From the initial and sudden white spots on her face, she was diagnosed with vitiligo.

She had been a beautiful child, a beautiful teenager. Her enlarged image, at fifteen, occupied the window of a local hair salon, an informal advertisement. Clara had fetishized our mother's appearance, her thick hair and high cheekbones, green eyes and full lips and flawless skin; she had tended and safeguarded and augmented; and now, in a matter of months, all of Clara's pride was converted to panic, all of her devotion went into concealment. With what little money she had, Clara sought out special makeup, but there was no way to completely hide the spreading areas that were devoid of pigment, the mini-continents and archipelagoes of unmitigated white on our mother's cheeks and forehead and neck, a white that contrasted dramatically with the rest of her complexion. Behind the cosmetics, pale clouds floated visibly across her face. Beneath her clothes, the stark, shapeless splotches overtook her body.

She underwent this transformation at sixteen, and when Bob and I were young, in our Brooklyn apartment, where the bathroom and master bedroom were separated by a long hallway, all of her mortification—and all of her mother's shame—could be heard in a scream she let out whenever she walked from bathroom to bedroom after a shower. She screamed out to our father, screamed for him to confirm that we were in our rooms with our doors shut and that he was at the other end of the apartment, that no one would see her without her makeup and with her body somehow insufficiently hidden by her full-length robe.

The layout of our home in Seattle didn't require this. But the intensity that went into the purchasing and private application of her makeup created a vibration throughout the house. She remained beautiful, when the cosmetics were aided by generous light or distance or a cursory glance. I remember that once, when we were driving, a group of college athletes catcalled her from the window of a team bus. I remember a piano tuner flirting. Yet at the same time, she was marred, and the gap between the fiction of her public face with the pallid, subtle clouds and the fact of her disfigurement must have felt so wide that she was always struggling, in near futility, to keep herself from plummeting, as if she stood in the middle with little or nothing beneath her. She had a doctorate, a rare achievement for a woman of her generation. She had been hired as an associate professor of sociology at the University of Washington; her research was gaining traction. But this didn't undo what had happened to her at sixteen. It didn't lessen the need to conceal, a compulsion that left her not enough energy for other forms of self-awareness—for checking her impulses, for balancing her love between her sons.

So my brother was alone and intimidated at school, detached in his bedroom at home, scared of our father, diminished by our mother. And yet, threatened by him—by everything from his lighter, straighter hair and handsome features to the nimbleness of his limbs to the knowledge that there was no difference in the powers of our intellects—I pinned and kneed and punched and humiliated him, until, on one occasion, our father intervened in the most hapless way, swatting my back with a folded section of newspaper, rather than pulling me forcefully off, which he could have done, with a brutality to meet my own. Or I stood outside Bob's locked bedroom door, with him inside, with our parents off at work, took off my belt, and began whipping the door, first cracking the leather end against the wood and then switching, so that the metal buckle struck the door, scraping the paint, then, harder, chipping the wood, denting it, making it splinter, whipping and daring him to come out, whipping and whipping.

Maybe it's no wonder that some months before my brother was put on a psych ward, and put on an aggressive regimen of Haldol, he took to walking around Seattle while wearing a tee shirt he'd had made, a shirt with a confession printed across his chest, a soundless cry for acquaintances and strangers to read: "I feel like I don't have anything to offer."

FOUR

Haldol's older, less-potent cousin, Thorazine, was psychiatry's first pharmaceutical breakthrough. It had its psychotropic debut in France, in 1952, in a trial with one subject, a laborer who'd been hospitalized for standing up and giving zealous political speeches in Paris cafés. He had also been hectoring people on the street. The drug calmed him right down. Next, Thorazine was given to dozens of patients on a Paris psych ward, where the psychiatrists praised it as "hibernation therapy." Once treated with the drug, they reported, the typical patient "remains silent most of the time. If questioned, he responds after a delay, slowly, in an indifferent monotone, expressing himself with few words and quickly becoming mute. . . . He does not express his preoccupations, desires, or preference." In the United States, psychiatrists welcomed the medication after watching what it did for the psychotic and the manic. It made them "waxlike, quiet, relaxed, and emotionally indifferent." In Canada, the drug was marketed to psychiatrists as providing a chemical lobotomy.

This was the outset of modern psychopharmacology. Behind

this beginning, Thorazine had a multi-use history. A chemical relative was a fabric dye in the Victorian era. Then, through the mix of flukes and determined tinkering that has sometimes been essential in medical discoveries—with penicillin as the prime example—the dye was turned into an antimalarial drug, and next, a cousin of the antimalarial became a tranquilizer used before surgeries. This tranquilizer soon found its way into psychiatry, as an antipsychotic.

There were early warnings that beyond subduing the delusional and unruly, the drug, taken regularly, might be causing havoc in the brain and throughout the nervous system. The director of a Swiss asylum tried the medication on three hundred patients and reported, in 1954, that over one hundred were quickly suffering symptoms of Parkinson's disease, a neurological disorder that destroys motor control. American psychiatrists identified some degree of Parkinsonism in nearly all who took the drug. There was shuffling. There was facial rigidity and drooling. In addition, patients fell prey to interminable twitching and spasms. Whatever Thorazine's benefits, they came at a clear neurological cost. No one knew how the drug worked in the brain. At a conference in Philadelphia, in 1955, psychiatrists debated whether the damage to motor control should be viewed not as a side effect but as somehow vital to treatment. Some argued that Thorazine doses should be pushed higher, to guarantee severe impact on the motor system and thus enhance psychological outcomes.

If this sounds like professional madness, still it's worth trying to grasp how such crude logic and overall enthusiasm was possible. The inventor of the lobotomy had won his Nobel only six years before that Philadelphia conference. Here, in Thorazine, was a more sophisticated, more medical alternative, a means, without driving a pick through the eye socket and into the brain, of achieving similar psychological results.

And then, too, Thorazine arrived during an era that saw a

popular embrace of lesser tranquilizers. While Thorazine was considered a major tranquilizer and offered heavy-duty sedation for the psychotic, U.S. pharmaceutical companies brought to market, starting in 1955, pills it called minor tranquilizers and pitched to mainstream America as a cure for everyday nerves and nagging distress. The pills were named Miltown, Equanil, Harmonyl—the forerunners of Valium—and the pitching wasn't done directly. The industry wasn't yet allowed to advertise prescription drugs in the lay press or on TV; that would come later. But it could promote directly to doctors, which became easier in the 1950s, when the American Medical Association dropped its requirement that drug companies provide evidence for the claims they made in ads that ran in AMA-published journals. This led to a flood of industry advertising, and as a result, the leading professional organization in American medicine more than doubled its revenues within a decade.

All of this meant that, through messages aimed not only at psychiatrists but at family physicians, and passed along to patients, drugmakers could shift the public's expectations about how a person should feel. An Eli Lilly brochure for its minor tranquilizer Ultran whispered that even grieving for the death of a loved one needn't be too painful. The brochure featured a mother and two children standing at a grave, with the mother gazing toward floating Ultran pills, below which was a simple, soothing promise: "attractive turquoise-and-white pulvules of 300mg., in bottles of 100."

Wyeth, makers of Equanil, quoted Macbeth in its promotions:

Canst thou not . . .
Raze out the written troubles of the brain,
And with some sweet oblivious antidote
Cleanse the stuffed bosom of that perilous stuff
Which weighs upon the heart?

With Equanil, eternal, Shakespearean woes were unnecessary.

Pfizer added to the list of troubles that, with its tranquilizer, need not be experienced: homesickness in children, "nervousness" in young men "caused by initial encounters with the business world," and, for all humanity, "familial tension." To help make its point, Pfizer mailed slippers and pillows, embroidered with the tranquilizer's name, to one hundred thousand physicians, an innocent- and insignificant-sounding gesture, except that at the same time the company, like its competitors, was submitting dubious studies of safety and efficacy, seemingly written by independent researchers, to journals that welcomed the industry's advertising dollars.

"Rarely is publication refused," a former Pfizer executive, recently resigned from the company, would soon tell a Senate subcommittee looking into the prescription drug industry, an investigation that would have negligible effect on how drugs were marketed. "Much that is called research in the pharmaceutical industry has little relationship to what most people engaged in academic and research activities would consider to be scientific research." The former executive continued: "In addition to the constant stream of promotion applied directly to the physician, there is a rather intense effort made to reach him through the patient"—that is, through patient requests motivated by glowing stories in the media, where "much of what appears has in essence been placed by the public relations staffs of the pharmaceutical firms." In the end, "the patient . . . is often exposed to drugs which have been incompletely evaluated, and which not infrequently are hazardous."

The company that made Miltown commissioned Salvador Dalí to design an interactive exhibit for an AMA convention: a gigantic walk-through caterpillar of anxiety; a butterfly of calm. The slippers and pillows, caterpillar and butterfly, journal articles and mainstream coverage, all played their parts in the messaging that spread throughout the country. Inevitably, too, some of the industry's work was done for free. Milton Berle, Mr. Television, host of a hit variety show, joked fondly about his regular use of Miltown pills, calling himself Miltown Berle. "You don't want a girl," he told Elvis on the air, in 1956, "you want a Miltown." Drugstores ran out of the medication and posted exultant signs in their windows when they got more in: "Yes, We Have Miltown!" Bars offered Miltinis, with a pill in place of an olive. For a year or so, until the others caught up, one-third of all prescriptions written in the United States were for Miltown. It was, then, the fastest-selling medication in American history.

The culture had come to believe that everyone should be one pill nearer to tranquility. Reports about addiction to the drugs cut into sales—but only briefly. And perhaps the nationwide quest for medicalized calm helps to explain the celebration of Thorazine among psychiatrists. There was shuffling, yes, and drooling, yes, but their patients—"indifferent" and in "hibernation"—could be viewed as serene.

Yet the company, Smith, Kline, and French, that owned the American rights to Thorazine wasn't taking any chances. Needing to be sure that there would be no backlash against the drug because of its side effects, it began a public relations campaign. Jointly with the AMA, it produced a new national TV show, *The March of Medicine*. After the opening program about progress in battling heart disease, the show turned to trumpeting Thorazine. The drug had been "administered to well over five thousand animals," the company's president told viewers, "and proved active and safe for human administration. We then placed the compound in the hands of physicians in our great American medical centers. . . . The development of a new medicine is difficult and costly, but it is a job our industry is privileged to perform." In fact, industry expenditures on promotion dwarfed research budgets by around four to one.

Smith, Kline, and French—which *Fortune* magazine ranked as

the second-most efficient profit-maker in all of American industry in 1958, and which would, after adjusting for inflation, quintuple its annual revenues between 1953 and 1970, largely because of Thorazine—did more than brandish its ethical and expensive investment in research. It contributed to something verging on the surreal. The Parkinsonism and other motor control effects seemed to disappear. In the most prominent press coverage of the drug during its early years, side effects were scarcely mentioned. The *New York Times* listed only trivial drawbacks—"Nasal congestion, temporary drowsiness, occasionally bizarre dreams"—in a story declaring that Thorazine would "revolutionize the treatment of certain mental illnesses." The writer touted the drug's "humanitarian and social significance." He remarked on the public funds that would be saved as veterans were freed from psychiatric disabilities. "There is probably no better example today," he concluded, "of the fact that research pays off in both lives and dollars."

In another *Times* story about the drug, a reporter toured an asylum where the medicated patients had, according to the staff, "regained their desire to fit into the world" and where many were swiftly well enough to be sent home. The story contained no interviews with patients. *Time* magazine compared Thorazine to the "germ-killing sulfas" that transformed medicine in the 1930s. But the most surreal rendering came in a study carried out, in 1961, by the government-run NIMH. Two hundred and seventy psychotic subjects were treated with either Thorazine or one of two recently released, closely related medications. The study found that after six weeks, ninety-five percent of the medicated subjects had improved, and that almost fifty percent were now "normal" or merely "borderline ill," with a further twenty percent only "mildly ill." The researchers stated: "The findings of this study lend strong support to the rising optimism about and confidence in the effectiveness of treatment of acute schizophrenic psychosis." They urged long-term maintenance on the drugs for fear of relapse. And they evaluated the medications' side effects as "mild and infrequent . . . more a matter of patient comfort than of medical safety." All three drugs had done wondrously well, and with all three, worrisome side effects were somehow unseen.

A kind of magic had occurred. The indifference noted as common in previous Thorazine patients and the loss of motor control discussed at the Philadelphia conference were not only gone; the researchers wrote that the drugs actually reduced apathy and improved movement. An alternate reality had taken seemingly objective, scientific form. Had public relations efforts behind Thorazine and the other drugs worked their way so thoroughly into the minds of the researchers as to alter perception? Was the vanishing of side effects a matter of psychiatric wishful thinking?

The side effects were definitely real and egregious enough— that was made plain by the truth that lots of patients avoided taking the pills. The industry came up with options like long-acting injectables for those who subverted treatment. "Warning!" read an ad in journals from Smith, Kline, and French. "Mental patients are notorious DRUG EVADERS. . . . Unless this practice is stopped, they deprive themselves of opportunities for improvement or remission." And they "deceive their doctors into thinking that their drugs have failed."

Greed and hope are great collaborators. The industry must have cheered when President Kennedy, in 1963, eight months before his assassination, announced to Congress that "new knowledge and new drugs" were about to swing wide the doors of psychiatric wards. Based on these medical advances, he called for a nationwide program of deinstitutionalization and community clinic-building. His proposal also addressed the educational and vocational needs of those then labeled retarded. He was haunted by the story of his intellectually disabled, lobotomized sister, long confined, concealed, and unvisited in a midwestern institution. Whether the issue was mental illness or low IQ, he was voicing the hopes of so many families of the afflicted. The new antipsychotic medications, he said, would replace "the cold mercy of custodial isolation." They would enable "the mentally ill to be successfully and quickly treated in their own communities and returned to a useful place in society."

The president signed the program into law weeks before his death. The "scientific achievement," he said, of "a nation as rich in human and material resources as ours" was making "the remote reaches of the mind accessible."

CAROLINE'S SISTERS LOVED to hear her read aloud. They were three and six years younger, and, in the back of the family car or on the living room couch, they pushed books at her chest and chin. On the couch, one sister got her lap and the other turned the pages. They were fans of a children's series, ValueTales. There was *The Value of Helping: The Story of Harriet Tubman*, and *The Value of Laughter: The Story of Lucille Ball*, and *The Value of Determination: The Story of Helen Keller*, with Helen sniffing roses on the cover. These were constantly requested, but the sisters were enthralled, too, when Caroline read to them from one of her adult books, though they had only a dim understanding of the content. What they adored, what kept them transfixed no matter the material, was the way Caroline changed her voice during the dialogue and whenever the narration shifted from one character to another. She had distinct tones and pronunciations, different rhythms, volumes, and moods for everyone. Each alteration sent a current of amazement through her audience.

This talent drew from the voices she heard. It wasn't that she gave those voices to the characters in the books but more that her mind was well tuned to the nuances of speech, because she heard speech so intimately and at such close proximity and so relentlessly, and this gave her the ability to construct particular sounds for the people in the worlds of the pages. For her, reading aloud, whether to her sisters or herself, was a necessary pleasure; it displaced and quelled, somewhat, the people who existed neither within the books nor within the parameters that her family and other human beings inhabited. For a while, when she was quite young, she didn't grasp that her companions were unheard by and unreal to others, and the realization of this disparity in perception, that she lived with people others could not perceive, never entirely took hold as she grew up. She knew it; she came to believe it; she accepted it; yet it wasn't fully internalized. The people in the unshared world were too present. She didn't see them, only heard or at times felt them, but they were there, the words coming not from within but from precise places outside herself: positioned to her side as she sat in a room; walking behind her as she walked.

The man who warned her about the daycare ladies, the impending nearness of the Gulf War, her parents' coming deaths, said, "We have to do something. We have to do something." Another, whom she knew as Miss Kathy, castigated her, telling her that her clothes smelled, that she was dirty, that she'd better keep still. Once, a teacher told Caroline's mother that she was sure Caroline knew the answers in class and that she needed to raise her hand and speak up. Miss Kathy asked, "Who do you think you are?" and Caroline kept her hand down and her mouth shut. One voice cried inarticulately. Another said, "We're nothing but shit beneath their shoes." Sometimes they were all reduced to a wall of sound, like an amplified version of the blurred voices of a crowd. Sometimes she couldn't tell if their words were directed at her or one another or someone others would recognize as in the room. Sometimes a voice was so frightened and loud, so far beyond constraint, that she would stay in her bedroom lest it erupt into a scream that would deafen her and be heard throughout the house.

Her family made a trip to Washington, D.C., and visited the Vietnam Memorial. She had only a hazy idea about what the black wall of names stood for, yet one of her voices became upset, and as they all—she, her family, this voice—descended the slope, the wall rising, the names accumulating, the two of them sobbed, baffling and disturbing her parents, who had explained the memorial only in the most general way. This was just before she permitted her parents a minimal hint of all she heard, just before she began seeing a psychiatrist and swallowing the medicine on the saucer set out with her breakfast each morning.

Her father tried to make light of the situation, commenting to her mother, in Caroline's presence, that she was "dancing to music no one else can hear." Her parents, both lawyers, tried to hide their panic, like they tried to hide the book she discovered, on top of the refrigerator, about how to raise a mentally disturbed child. "They're going to kill us," a voice said, the "they" unidentified. A narrator accompanied her throughout her days, his tone always authoritative and sometimes faintly mocking: "She's getting out of bed now. Oh, she's walking down the hall now."

The pills she was given, by the time she was in middle school, included what psychiatry termed second-generation antipsychotics. The first generation had started with Thorazine and led to Haldol. The second—clozapine, Risperdal, Zyprexa, Seroquel, Abilify— was introduced, in the late 1980s and into the '90s, as a leap ahead. This new category of medication would not only curb psychotic hallucinations like Caroline's voices, it would treat other symptoms that sometimes came along with psychosis, symptoms that the earlier antipsychotics didn't touch, such as particular patterns of social withdrawal and cognitive dysfunction that seemed possibly distinct from what could be attributed to voices, presences, paranoias. Perhaps the best improvement seemed to involve side effects, a problem that had been gradually acknowledged with the first-generation drugs. The second-generation pills, the industry asserted and practitioners believed, spared patients these disabilities.

For Caroline, though, the second-generation drugs brought no reduction in her voices, regardless of the combinations her psychiatrist tried, combinations of chemicals that were very similar, so that the process was more of a piling on than a meticulous putting together. What Caroline learned, throughout her first several years of treatment, went deeper than the idea that her voices were a manifestation of illness and that they must be silenced. She learned that she was uniquely sick among the ill, because the drugs weren't working.

As for side effects, no one spoke with her about them, at least not as drug-related. She was fifty pounds overweight because she ate too much and didn't get enough exercise. The quaking of her hands and sporadic twitching or jerking of her neck and head went unmentioned by the adults around her. It was as if the spasmodic movements were nightmarish imaginings, delusions of her diseased brain, except that the taunting at school confirmed their reality. And because no one said that these were side effects, that all her physiology, from her metabolism to the control of her fingers, was being affected by the medications she was swallowing, she assumed that not just the obesity but the quivering was a hideous flaw of her own making.

"You disgust me," the girls at school said. She disgusted herself. Anxiety spiked with their revulsion and worsened the trembling and, too, the twisting of strands and clusters of hair around her fingers, the yanking out of hair at the roots, exposing her scalp, exacerbating the disgust.

Home at the end of her day, in middle and high school, she lay on her bed for hours until her mother roused her for dinner. On her walls were posters of Kurt Cobain and the feminist punk band Bikini Kill. There was Alex, leader of the incorrigible and savage droogs in *A Clockwork Orange*. She lay, not exactly sleeping but semi-comatose, needing to escape the school day she'd just endured, needing to recover from the fatigue of that endurance, and exhausted from the antipsychotics that were supposed to make her

normal. Homework was impossible, though a stimulant, an attention deficit disorder drug to help her concentrate and get her work
done, was among her mix of pills. Her grades were dismal.

The one thing she had focus and energy for was reading. She
pored through Bellow's *Herzog*, as its narrator churned out imagi-
nary letters to the famous and the dead. In science class, holding
the book open under her desk, she lost herself in Sinclair's protest
novel about the plight of immigrants, *The Jungle*.

"She's leaving her room now," her narrator said. "She's going to walk to the bathroom. She's looking at a razor blade."
Around the time that Cobain killed himself, she started cutting.
In this, there was control. There was relief. It was a course of
action she could decide on and carry out. Its result, in blood and
pain and damage, briefly purged her. She used nail scissors as
well as razor blades. Once, in her rush to attain this liberation,
she used the blunt side of a single-edged blade, pressing hard to
penetrate the flesh and drawing the steel across, opening a wide,
jagged gash.

Cobain's suicide, by a shotgun blast to the head, wasn't the
reason for the cutting. She wasn't distraught over his death,
though she was devoted to his and Nirvana's music, overcome by
it, feeling that his songs were written for her. His harsh lament,
"Been a Son," spoke to her certainty that she should have been
born something other than whatever she was. She was hypnotized
by the throb that opened "Heart-Shaped Box," by the thrashing
that followed, by the yearning for submission:

She eyes me like a Pisces when I am weak

and the violent longing for a primal, prenatal innocence:

Throw down your umbilical noose so I can climb right back.

When he died, she was inspired to draw a cartoon. She intended it to be in the quiet, wry style of cartoons in the *New Yorker*, to which her father had a subscription. She sketched two girls talking about how horrible school is. Then she drew Cobain. He materialized before them, shotgun in hand. "Well," he said, "you can always try this." She stared at what she'd made. She had aimed for subtlety and saw that she had accomplished no such thing. Yet there was something else unforeseen. A recognition began to form. It was incomplete, but it lurked in her inept cartoon. It was that living or dying was a choice, that both were voluntary, that death was an option. This was jarring but not distressing. It was comforting.

"Oh, I don't think she's getting a ride to the truck stop this morning."

She had, at fifteen, sixteen, a few friends—or, at any rate, other misfit smokers and druggies willing to coexist with her. One ritual was to have breakfast at a truck stop on the highway toward Chicago, ensconced by the giant rigs in the lot, and get high before school.

"No, no, no. They're not coming to pick her up. Caroline's not going. Not going anywhere."

They were dependable enough, though, when it came to buying weed, Valium, Ecstasy, heroin. Zionsville had ample supplies. She knew she wasn't supposed to be mixing street drugs with her prescription medications, but this didn't give her the slightest pause. Her only worry was a rumor that one of the drugs in her psychiatrist's mixture, an antidepressant, could dampen the effects of Ecstasy.

One afternoon, she brought the narcotics she'd just purchased home with her: heroin and Valium. It wasn't unusual for her to do drugs by herself at home, and it wasn't unusual for her to do heroin and Valium together. She snorted the powder and swallowed the pills.

But she did have a sense that this time she might have done too much. She'd chosen a mostly unused room in the house. She didn't know why she wasn't in her bedroom with the posters of Cobain, Bikini Kill, Alex the droog. Maintained by her mother, this room served as a kind of museum. It preserved the past. Years ago, Caroline and one of her sisters had slept here together. The walls remained pink. There was the white wicker dresser the girls had shared. There was the white wicker trundle bed where they had slept. Spread smoothly over the bed was the same pink cover adorned with a topography of rosebuds. There were the same ruffly pink pillows.

Caroline lay back on the rosebuds after ingesting the drugs. No one else was home. She thought: *That was probably more than I should have done. . . .* And a while later: *This may be it for me.* The thoughts were not alarming. Her calm was unrelated to her fledgling realization as she had stared at her cartoon, the idea that death could be a decision, something she could opt for or against. This was different. This was passive, not a decision at all. It felt like both a minor error and a major inevitability. She was aware of drifting out of consciousness. She thought: *I might not wake up.* This caused no regret, no effort to avert what was happening, no impulse to interrupt the fading out. All was apathy.

She did wake up, abandoned by the haven of unconsciousness. She was a failure, too, at death. Groggy, she went outside and lowered herself into a hollow against the wall of the semi-basement. She smoked, her body compressed partway under the house.

A lesser haven, during these years, continued to be reading aloud. She had read *Animal Farm* to her sisters, conjuring the speech of the pigs and horses and the singing of "Beasts of England" before the animals took control of their own governance— and before the revolution was ruined by the pigs' lust for power and insistence that they knew what was best for all the other creatures. And lately, she read them *The Old Man and the Sea* and

The Moor's Last Sigh, inventing pitches, textures, and cadences for Hemingway's fisherman and Rushdie's Moor.

Her grades may have been dismal, but when she took her college entrance exams, she received a perfect score, putting her in the ninety-ninth percentile, on the verbal section. This was enough to compensate for the rest of her record. She was going to get a new start, away from her hometown, away from herself. Caroline was headed to Bloomington to attend the state university.

Science, President Kennedy was sure, would take us to "the remote reaches of the mind." And science, he vowed, would take us to the moon. By the end of the 1960s, we were walking on the moon's cratered surface. As for venturing into the mind and finding new ways to medicate the most troubling psychiatric conditions, we had progressed from Thorazine to Haldol, which was released in America in 1967.

The chemical was discovered by a Belgian pharmacologist whose company would soon be bought by Johnson & Johnson. The Belgian researcher had just developed an opioid painkiller and was looking for something even stronger. He employed a team of fifty chemists, most of them self-taught and low-paid, and directed them in a simple method: synthesize chemical offshoots of the original painkiller as quickly as possible, inject the new molecules into mice, and see what happens. One of the molecules—the forty-fifth in the series they concocted—had a curious effect. It didn't seem to work as a painkiller. This the researcher could tell by measuring the pupils of his rodents when they were placed on a hot metal plate. But while the animals did register pain, they seemed paradoxically unconcerned about it. The chemical put them in a trancelike state, apparently unmotivated to lift their paws. Second after second went by—thirty seconds or more ticked past—and the mice stayed where they were on the scorching surface.

The chemical intrigued the pharmacologist. He gave it to Belgian psychiatrists for trial-and-error experiments. It was injected into two alcoholics going through the hallucinations of delirium tremens and then into eighteen psych ward patients. "We consider its hallucinolytic action"—its reduction of voices and visions—"to be greater than that of any other neuroleptic," the psychiatrists wrote, referring to Thorazine and other antipsychotics, though they did note that with the new chemical, "Parkinsonism is the norm." Next, one of the French psychiatrists who developed Thorazine tested the new drug. He affirmed its effect on hallucinations but warned about the highly prevalent damage. In a French journal, he urged that dosages be kept moderate to lessen the side effects.

Psychiatrists now had, it seemed, a better molecule to pit against psychosis, if it was used judiciously. Gradually, though, it was not only the drug's safety but its efficacy that was questioned. Much later, *The Lancet*, Britain's premier medical journal, published an analysis spanning backward over more than half a century's worth of studies assessing antipsychotics: Haldol didn't do impressively well—even when side effects were left out and the calculation included only addressing psychotic symptoms. And in America, a judicious concern for Haldol's side effects never did prevail. Before Haldol, when Thorazine was the star antipsychotic, Nathan Kline—a luminary in American biological psychiatry, who had won two Lasker Awards for public service in medicine and whose name would grace an organization, the Nathan S. Kline Institute for Psychiatric Research, that Goff now directed in addition to his position at NYU—advised the field that only "massive doses" would bring about successful outcomes. U.S. practitioners heeded his advice on Thorazine and kept right on once Haldol took over. A study published in the *American Journal of Public Health* showed that typical doses doubled between 1973 and 1985, climbing to levels many times higher than in Europe.

The reasons for this aren't easy to pin down. Professional or-
ganizations did their part. Right after speaking on *Oprah* about
the risks posed by Haldol, one psychiatrist found himself under
investigation, fending off an attempt to revoke his medical license.
And psychiatrists likely managed to deceive themselves. A potent
medicine is a seductive thing to possess and prescribe. But neither
practitioner self-deception nor corporate self-interest is limited to
America; they are insufficient explanations for American dosages.
Perhaps, then, American culture is particularly prone to a faith in
cures and a belief that aggressive methods will prove to be remedies,
even panaceas.

Goff saw the reality of Haldol fairly early, from up close. As
an undergraduate in the 1970s, he had studied philosophy and
dreamt of becoming a novelist but steered toward medical school.
"I caught the wave of biological psychiatry in my residency," he
remembered, thinking back to the early '80s. "It was an exciting
time." Yet he was enamored more of biology's future, of drugs
yet to be discovered, than of the medications he could then prescribe. He did a stint at a Boston psychiatric clinic, and every
day, as he walked to work and came within a few blocks, he saw
the harm done by Haldol. It was embodied in front of him in the
people he passed on the sidewalk; it was impossible to ignore. "As
you approached, there were the patients from the clinic with their
strange movements, their bent-over bodies, their tremors. From
the meds. Not only was the illness debilitating; the meds left them
so physically miserable."

Still, a general optimism had been building that the brain
would, at last, reveal its secrets, that the mystery of its machinery
would be solved, that revolutionary medications would follow.
Or maybe it's more accurate to say that even during the period
when medical psychiatry felt itself under existential threat, such
optimism never actually abated; maybe the materialists of the
mind inherently trust that the psyche's problems are waiting to

be unraveled and cured, because machinery, no matter how complex, is ultimately finite in its intricacy.

"For a decade and more, research psychiatrists have been working quietly in laboratories, dissecting the brains of mice and men and teasing out the chemical formulas that unlock the secrets of the mind," the Baltimore *Sun* declaimed in a series that won the Pulitzer Prize for explanatory journalism in 1984. "Now, in the 1980s, their work is paying off. . . . Psychiatry today stands on the threshold of becoming an exact science. . . . Ahead lies an era of psychic engineering, and the development of specialized drugs and therapies to heal sick minds. . . . The mind-brain barrier has been probed and found not to exist." The writer then quoted a neuroanatomist, "'The brain is an organ; it produces thoughts the same way the kidney produces urine,'" and another neuroscientist, a senior researcher at the NIMH, "'People who act crazy are acting that way because they have too much or too little of some chemicals that are in their brains. It's just physical illness! The brain is a physical thing!'"

The story continued: "There are great principles to be discovered, Nobel prizes to be won, fortunes to be made with new drugs. And, day by day, the precise chemical nature of human thought and emotion grows ever more clear."

This "chemical nature" boiled down to neurotransmitters, molecules that link up the brain's hundred billion neurons and the body's nerve cells as well, molecules whose scientific history starts with experiments a century ago. Once, neurologists believed that the nervous system did its signaling with electricity. But then, in a dream, in 1921, a scientist saw a frog experiment fully designed before him. He woke up and went about discovering that he could control the speed of a frog's heartbeat by using either electricity or a fluid from within the frog, a fluid that contained acetylcholine, the first neurotransmitter to be isolated. Both electrical impulses and chemical substances were somehow involved in signaling. Yet it was assumed, for decades, that this was the case in the body but not within the brain.

Lysergic acid diethylamide, LSD, helped scientists to see that chemicals are essential to communication between brain cells. The drug had been synthesized by a chemist hoping to find a cure for migraine headaches. He absorbed some through his fingertips and was engulfed in fantastic shapes and colors, then swallowed a sizeable amount and was surrounded, he reported, by furniture that took on "grotesque, threatening forms" and a neighbor who became a "malevolent, insidious witch." Meanwhile—still in the body and not the brain—a new neurotransmitter had been discovered, serotonin. These lines of exploration converged, in the 1950s, when researchers focused on two facts: that the chemical structure of LSD resembles the composition of serotonin, and that serotonin had been found in animal brains. If signaling in the brain depended on serotonin, it would make sense that a structural cousin, LSD, mistaken by neurons to be the real thing, could block serotonin's work, causing the brain to go haywire.

Thwarted serotonin didn't turn out to be a straightforward explanation for LSD's cerebral chaos, but scientists were now drawn to the brain's dependence on chemicals, on neurotransmitters. Between each of the brain's neurons—between the "little sorts of tails" that Dostoyevsky wrote about, roughly referring to the tendrils that extend from the cell bodies—lie synapses, gaps, microscopic clefts. Systems of neurons in the brain need to send messages, neuron to neuron, along elaborate tracks within and between brain regions. To do this, they need a way to bridge the synapses, which number well over a hundred trillion. And this bridging needs to be temporary, instantaneously brief, so that the systems operate with modulation and avoid overdrive. The neurotransmitters, released by one neuron and acting deftly upon the next, do the bridging.

The discovery of yet another neurotransmitter, dopamine, came slightly behind the initial brain research on serotonin. And when scientists tranquilized mice and then injected their brains with a dopamine precursor, they watched the torpid rodents perk back up. Soon, dissecting the brains of deceased Parkinson's patients, researchers found atrophy in a brain area that is a crucial supplier of dopamine to a region that governs movement. Deficient dopamine was thus pinpointed as the crux of an immobilizing disease. By the '70s, scientists knew that Thorazine and Haldol inhibited dopamine, that this was why the drugs caused such physical wreckage. And researchers began to investigate an idea: that the medications had their somewhat ameliorative effect on psychosis because hallucinations and other unrealities were the product of excess dopamine in other areas of the brain, areas that shape perception and thought.

The abuse of amphetamines during the 1960s and '70s offered indirect confirmation for the theory that dopamine was the culprit in psychosis. Amphetamines boost dopamine along cerebral pathways. The speed freaks of the era were appearing in hospital emergency rooms with usually temporary delusions; here was suggestive evidence that dopamine was to blame in the psychiatric disorder. In 1973, the neurologist Oliver Sacks published *Awakenings*, his bestselling account of working on a ward of Parkinsonian patients, victims of an encephalitis outbreak who had been more or less paralyzed for decades. He treated them with the dopamine precursor levodopa. The resulting surges of dopamine enabled patients to play the piano and walk through hospital gardens, until the levodopa strangely lost its power and the patients returned, unrescuably, to paralysis. Along the way, something happened to some of their minds. The levodopa did more than liberate bodies; it led to wild beliefs. One patient pleaded to be sent on a lecture tour. The world was teeming with devils, and he, the messiah, had been called to combat them. Here was another sign that psychosis could be attributed to too much dopamine.

Any lucid explication written about the brain's neurotransmitters risks simplification to the point of being misleading, because the brain defies lucidity. But incrementally, from its beginnings several decades ago, research has outlined three main dopamine pathways. They originate just above the brain stem, near the top of the spinal cord. One runs to an area that controls movement. Another channels into the limbic regions—among them the hippocampus, the seahorse-shaped zone Goff dwelled on—that influence attention, perception, fear, desire, memory. A third engages the frontal areas, the province of complex cognition. Dopamine systems reach throughout the organ and facilitate most of what we feel and do.

A climactic part of the Pulitzer-winning Baltimore *Sun* series heralded the effort to find a drug that would treat psychosis by depleting dopamine along the paths where it seemed to need diminishing, while leaving the neurotransmitter intact where depletion did harm. The drug would be finely targeted. It would go a great way toward eliminating side effects. The relevant science was racing forward. "We've gone from ignorance to almost a surfeit of knowledge in only ten years," the story quoted a mentor of Goff's.

Goff himself, despite the "physically miserable" patients at his Boston clinic, sensed "the possibility of limitless progress," he recalled. "Those were heady years." One medication was already affirming hope. "Clozapine came along and brought feelings of huge promise. It not only stopped psychosis but all the Parkinsonian side effects were gone. There was weight gain but none of the movement side effects." The problem with clozapine was that it could kill patients by destroying their immune systems, so its use was permitted only as a last resort and only with close monitoring. And, too, no one quite knew how the drug worked; it didn't define a clear route for new research. The weight gain was more than a moderate issue; patients were fast putting on fifty pounds and more. There were also side effects that Goff didn't seem to have focused on: hypersalivation, as researchers called it, in at least thirty percent of patients, and tardive dyskinesia in five percent. Five percent was a lot better than estimates that made motor control disorders the norm with Haldol, but clozapine wasn't bypassing motor control circuits.

Drug companies, though, seized on the potential in clozapine's improvements; they set out to build on whatever they could discern about the drug's mechanics. This led to the second-generation antipsychotics, which, in the broadest terms, tweak serotonin while they dial down dopamine, though no one has truly sorted out how the medications operate. Because Goff soon became the director of the Boston clinic, as well as being on the Harvard Medical School faculty, he was contracted by Eli Lilly to oversee some of the preliminary testing of its new antipsychotic, Zyprexa. "There's a saying in medicine that your opinion of a drug is formed disproportionately by the first patient you treat with it. We were running a study of Zyprexa with patients who hadn't done well on other meds, and I had a lovely patient, a woman in her forties, who was very psychotic. She had unremitting hallucinations. She also had scoliosis. This was separate from the psychiatric disease, separate from any side effects. The scoliosis meant that she could barely walk. She used arm crutches; it was painful. Then she got the Zyprexa. She did so well in the trial that afterward we put her on the drug, open label. Her psychosis was gone. But it wasn't just that. Suddenly she could walk without the crutches. I felt like the pope." Goff laughed, as he often did, at himself—and at the mystery. How the drug had freed this woman from crutches he would never know, beyond speculating that the energy she'd been exhausting in dealing with her voices could now be spent on getting around.

The second-generation antipsychotics became widely available in the 1990s, and a fast-growing organization, the National

Alliance on Mental Illness, welcomed them ardently. NAMI had begun as a support group founded, in the '70s, by two mothers with adult schizophrenics living at home. Despair and anger bound the group together, anger because some Freudians had long traced the disorder to bad parenting and, above all, to bad mothers—mothers accused of veering between overprotection and hostility. Biological psychiatry dispensed with such blame, and NAMI, which grew to 125,000 members by 1990, adopted neurochemistry as its cause.

Beyond lobbying Congress for research funding, NAMI promoted the latest drugs. "The long-term disability of schizophrenia can come to an end," its executive director told the *Wall Street Journal* in a story headlined "Psychosis Drug from Eli Lilly Racks Up Impressive Gains." The story described the competition between Eli Lilly and Johnson & Johnson—between Zyprexa and Risperdal—to become "the world's first billion-dollar-a-year antipsychotic." A book NAMI published, *Breakthroughs in Antipsychotic Medication*, opened with photos of second-generation pills against pastel backgrounds and counseled that while older antipsychotics "correct brain chemistry by working on the dopamine system," the newer drugs "do a better job of balancing all the brain chemicals, including dopamine and serotonin." For the anguished family member or patient, this was just enough science to sound authoritative and reassuring. There was no need to acknowledge that behind the talk of balancing lay a void of understanding, mechanisms no one had penetrated. Better, instead, to predict a happy outcome: "Give the new medication plenty of time to do a good job!"

The industry showed its appreciation by donating millions to NAMI every year, and NAMI was plainspoken about its agenda in a nonprofit mission statement filed with the IRS in 2000: Providers and pharmaceutical companies "want to grow their markets" and "NAMI will cooperate with these entities to grow the market by

making persons aware of the issues involving severe brain disorders."
Desperation, perhaps far more than the donations, dissolved the
borders between corporate goals and nonprofit activism.

"There was so much optimism," Goff said. "We were sure we
were improving people's lives." But for him, after his elation over
the woman who shed her voices and her crutches, and for others
willing to peer beyond hope, suspicions seeped in. As Zyprexa
progressed from trials to FDA approval, and as Goff and his clinic
team prescribed it to more patients in the late '90s, they noticed
a pattern of both alarming weight gain and diabetes. He went to
Eli Lilly with his worries, but the company's experts deflected the
problem, insisting that the drug did nothing to the metabolism
and that the obesity and dangerous cases of diabetes weren't due
to the medication. Goff and colleagues at Harvard's Massachusetts General Hospital studied the drug's impact, confirming that
Zyprexa crippled insulin's ability to regulate blood sugar. The
company continued to deflect.

Then, in 2006, a lawyer who'd twice been locked on psych
wards in his twenties and thirties, and who, ever since, had been
trying cases to protect psychiatric patients from being forcibly
medicated, got hold of hundreds of Eli Lilly's internal records
and emails about Zyprexa. He gave the documents to the *New
York Times*. By 1999, the records demonstrated, the company had
known of Zyprexa patients gaining over one hundred pounds. It
had data showing that sixteen percent put on more than sixty-six
pounds. "Weight gain and possible hyperglycemia is a major threat
to the long-term success of this critically important molecule," the
Eli Lilly scientist in charge of Zyprexa wrote to colleagues. "Unless we come clean on this," an Eli Lilly official warned in 2000,
after the company hired diabetes specialists to assess the situation, "it could get much more serious than we might anticipate."
That same year, an internal study showed that Zyprexa raised the

risk of high blood sugar by three and a half times. In 2001, Eli Lilly fretted over marketing reports saying that psychiatrists were talking about the connection between Zyprexa and diabetes, yet the company decided against distributing educational materials on managing the endocrine disease, because this "would further build its association with Zyprexa."

Eli Lilly chose, rather, to focus its marketing on primary care physicians. It banked on their credulity when salespeople called the drug a "safe, gentle psychotropic." The company pushed the medication to nursing home doctors as a valuable sedative for controlling irascible residents, and it promoted Zyprexa as a stabilizer for kids struggling with disruptive behavior—populations for which the FDA hadn't approved the drug. It peddled to state-run mental health centers that serve the poor, as did other companies with their second-generation antipsychotics; by the mid-2000s, foster kids were prescribed the drugs at a rate five times higher than among other children, and between one-third and half of these prescriptions were for attention deficit/hyperactivity disorder, ADHD. This prescribing wasn't exactly illegal. Once a drug is approved for any condition, the law leaves other uses up to the judgment of physicians.

Why the FDA had approved Zyprexa at all is a riddle answered by the low standards set by first-generation antipsychotics, by subtle and knowing fraudulence in the research submitted by the pharmaceutical company, and by understaffing at the agency. During the evaluation process, one FDA reviewer wrote that, given Zyprexa's broad and poorly comprehended action in the brain, "no one should be surprised if, upon marketing, events of all kinds and severity not previously identified are reported." His uneasiness was echoed by other evaluators. Zyprexa was nevertheless waved through, and by 2005 it was Eli Lilly's bestselling drug, bringing in upward of four billion dollars a year. Lawsuits over the damage done by the drug and over the company's deceptions wound up costing the company around three billion, a minor fraction of its profits from the medication.

The parallel story of Risperdal features an autistic boy and a revered Harvard psychiatry professor, Joseph Biederman. Johnson & Johnson's patent on Haldol expired in 1986; Risperdal, its second-generation entry, was released in 1993, and the company, despite specific warnings from the FDA, decided to market the drug for use in the elderly and in children. It created an elder-care sales force of over one hundred to focus on nursing homes. It led conference rooms full of reps in wolf howls and sent them off to pediatricians' offices. It provided two million dollars to launch the Johnson & Johnson Center for Pediatric Psychopathology at Massachusetts General Hospital, which Biederman directed. The professor received over a million and a half dollars in personal payments from drug companies between 2000 and 2007, much of it from Johnson & Johnson. The center's aim, Biederman spelled out in a 2002 annual report to the company, was to "move forward the commercial goals of J&J" by training pediatricians and psychiatrists to diagnose childhood bipolar disorder and by alerting them to the "large group of children who might benefit from treatment with Risperdal." Early diagnosis and treatment, he wrote, would have the added merit of leading to lifelong customers—to the "chronic use of Risperdal from childhood through adulthood."

The company's campaign to reach pediatricians wasn't limited to targeting bipolar kids, and in 2002 it touched an eight-year-old boy in Thorsby, Alabama, who was both bright and autistic, and who was having trouble with tantrums at school and home. His mother, a factory worker, found a psychiatrist in Birmingham, who suggested Risperdal. The drug worked—to some degree—against the outbursts. The boy stayed on it. The psychiatrist had no idea that in a 2000 study of over two hundred and fifty boys done by Johnson & Johnson, nine percent developed a prolactin condition that generated breasts. These were not the seeming breasts of obesity; these were actual breasts. The company, which had suspected the problem for some time, kept its data to itself. Yet beyond the Harvard professor, the company paid, in 2002, nearly two million to leading psychiatrists to spread the word about the value of Risperdal for children and adolescents. And meanwhile, the Alabama boy was being teased by neighbor children when he went swimming. He refused to swim without a tee shirt. By the time he turned twelve, he had full, feminine-looking breasts that would grow to size DD. His mother understood what was happening to her son only when she saw a law firm's TV commercial searching for takers of Risperdal who suffered gynecomastia.

Lawsuits and government penalties eventually subtracted three billion or so from Johnson & Johnson's Risperdal profits, inflicting around the same mild financial pain felt by Eli Lilly. But for some specialists in psychosis, like Goff, doubts about the whole class of second-generation antipsychotics predated the litigation and the exposure of company tactics. The doubts, though, didn't override the assumption that the new drugs were, all in all, an improvement. The field was sure about that. By 2001, Goff was part of an NIMH study with fourteen hundred subjects across the country, research designed to find out whether the new medications were better, in efficacy or safety, than their badly flawed predecessors. When the study was published, in 2005, the answer was plain: no, they weren't, perhaps not even where they were supposed to offer their strongest advantage, in doing less harm to the neurosystems of movement. "There was shock," Goff said. "It was a resounding disappointment."

AT COLLEGE, IN Bloomington, Caroline took a neuroscience seminar. She figured, I'll learn why I am crazy. The answer wasn't forthcoming, but the course, on the evolution of the brain, fascinated her. Humans had become humans, human brains had become human brains, from beginnings going back to sea creatures that didn't even have brains but navigated life with scant nerve netting spread over their flimsy bodies. Flash forward hundreds of millions of years and the brains of mammals grew as the animals struggled in territory dominated by dinosaurs. Tens of millions of more years and the social life of primates led to more brain size and to branches and branches of interlinking neural networks.

Caroline gave a presentation about visual processing and how we see color. She led her classmates, step by step, from the retina back and downward through the brain to the thalamus and up and farther back to the fusiform gyrus and the lingual gyrus in the occipital lobe. Along the way, she taught that for all that science could outline, researchers really weren't even close to sorting out the mechanisms between the eyes and the perception of green and blue. This excited her. And this was for mere color, nothing like cognition. After the presentation, the professor paid her a compliment about how she'd taught things. She'd interwoven the cerebral mechanics with relatable metaphors; she'd slipped in some humor and got everyone laughing. "You have a gift," he said. He told her that his wife taught a seminar in science writing at the journalism school, that the class was by invitation only, that Caroline was a perfect candidate, that he could imagine her having a career with a major newspaper. He would make sure his wife was waiting for her to get in touch.

She was caught up, too, in a religion class that delved into Spinoza, who saw the empirical as a spectacular expression of spiritual forces. And all the while she was discovering Bloomington. Wearing a wig like Liza Minnelli's in *Cabaret* and a black hoodie with a picture of the riot grrrls of Bikini Kill, she went to punk concerts at dives that doubled as meeting spots for activists. She studied carbon footprints. She protested for bicyclists' rights. She volunteered with

Pages to Prisoners, packing up books with handwritten notes and shipping them to the incarcerated.

But she also bartered sex for street drugs. She had heard the lesson, over and over, that medication was helping her, and she extended this line of thinking to the narcotics she smoked, swallowed, snorted, injected. The bartering wasn't always conscious trading. It wasn't prostitution. It was blurrier than that but no softer: men in states of addiction and states of rage; she absorbing their anger, their brutality; a man battering her in the shower; she waking up in a costume of florid bruises.

She never did reach out to the neuroscientist's wife, the journalism professor. She drifted away and dropped out of college. She got pregnant. She couldn't be sure but thought the father was a crust punk and anarchist she loved, a man with whom she foraged for freshly discarded food in a dumpster in an alley behind a café. As she walked into the abortion clinic, the building was surrounded by protesters. They waved wire hangers and held placards: the cherubic faces of babies floating below the word "LIFE"; the luminous faces and half-formed hands of fetuses suspended in the womb.

"Murderer!" they yelled. "Murderer!" Inside, while she was on the table, a clinic attendant tried to calm her, holding her hand and saying, "Just tell me a little about something you enjoy." Caroline felt that the young woman in scrubs expected her to talk about a pet dog or music or partying. Instead, she leapt into the evolution of the brain and the ideas of her religion class and how awe-inspiring and sort of sacred our physical selves were and how much she liked learning about all of it. The attendant's surprise was naked. Caroline had a sensation of being seen. "That's just like me," the woman said. "That's just like me."

After the abortion, on Caroline's way out, there were the placards, the accusers, and by that night her voices were joined by an infant shrieking. Others were frantic and others lancing. Another told her he would remove her fingers "one by one by one." She deserved it, he said; it was unavoidable. He spoke so close she could almost feel his breath. In bed, she felt him take her hand and begin to cut.

"Get the fuck away from me," she threatened them all. Alcohol and drugs were as futile as the threats. "What the fuck are you saying? Get the fuck away." She raved on the streets, clutching a wire hanger. She wrestled with the police who came too close. They arrested her and locked her in jail.

It was her second arrest. The first had been for refusing a police order to back up during a rally against the oil industry. Her third was for theft. Besides sex, she paid for narcotics with stolen goods or prescription pills that were worth something on the street—her ADD drugs, her benzos. Once, rifling through her parents' closets in need of something to sell, she'd found a bag of knives and scissors, and realized that her parents had hid everything she could use to wound herself or her family. There was scarcely a jolt of shame before she went back to searching. The arrest was for stealing electronics from a CVS. Her parents hired the right lawyer, who struck a deal with the judge; Caroline would be sent off to a psychiatric hospital in lieu of a criminal sentence.

Outside Houston, on a suburban campus of short, neatly pruned trees and long, low buildings, the hospital was divided into units, each of them renamed by the patients to form a special family. The Professionals in Crisis unit was the burnt-out dad; the Eating Disorders unit was the perfectionist daughter; and the Hope unit was the crazy aunt living in the family basement with no hope whatsoever—and this was where Caroline was assigned. After intake, she could feel that it was not an auspicious placement, and her fellow Hope dwellers confirmed her guess. She was the right age for the Compass unit, where young adults found their way. Hope, where most everyone was middle-aged and up, was where patients went to rot.

The patients were allowed their street clothes, with adjustments: no belts, no shoelaces. Most wore pajama pants, sweatpants, bath slippers, slip-on sneakers, flip-flops. A staffer took the strings from Caroline's hoodies, but she could wear her black sweatshirts, the one bearing Bikini Kill, another with a woman holding a revolver above the words "You Can't Rape a .38." She felt vaguely comforted when she put this second one on, and the staff seemed to sense this and let it pass. She was permitted the black wig that concealed her bald patches, and she could have access to her black eyeliner and nail polish under the supervision of a large male staffer, to make sure that she didn't drink them. She had also arrived with a bottle of perfume. Her grandmother, her father's mother, always wore a scent that blended rose and vanilla, and Caroline liked to wear perfume as a reminder. The bottle was taken from her at intake, but she could use it with a staffer standing over her. She was on an every-fifteen-minute suicide check. It was round-the-clock. Throughout the night, four times an hour, a man aimed a flashlight at her as she lay in bed. All the Hope dwellers were on the same routine, except for those on every five minutes and those on the nonstop watch of one-on-ones.

In group therapy meetings, in a room with pale bare walls, a whiteboard, and a tall, fake plant, she kept silent, and in private sessions she kept her words to a minimum, but in the smoking cage, she yearned to take part. The cage was a semi-enclosed porch with bars running from floor to roof on two sides and a view of the trees that looked to Caroline like they'd been ordered from IKEA. There wasn't a gnarly root or a majestic trunk or an outstretched, crooked branch in sight. Staff didn't come onto the porch but kept vigil through a glass door to the adjoining day room. In the cage, she listened to the older patients of her unit laugh about the absurdities of life at the hospital: how they had to stay on the therapist's chosen topic during group sessions and weren't supposed to say anything triggering, which meant that they couldn't talk about anything that mattered; or how, inexplicably, Hope and Eating Disorders were combined for group therapy, which meant a mix of the woefully fat and the sickly skinny.

Caroline wanted to contribute to their banter, to mutter some comment that would spur their dark laughter. She felt something reassuring in them; though they might be twice or three times her age and relegated to what they had rechristened the No Hope program, they had endured a lot of trials, a lot of years. They smoked Camels and Marlboros and Kools, but she ventured in with American Spirits. They burned the longest. The staff let patients into the cage even if they didn't smoke, but she needed a cigarette in her fingers to make her feel less outcast as she lingered, listening, at the edges.

One day in the cage, a patient everyone called the Godfather, an Italian man who spoke with the intonations of a Hollywood gangster, named her the Black Hole. She knew right away that it wasn't mocking, that it had nothing to do with the taunting that had besieged her in school. The name was in honor of her wig, her hoodies, her eyeliner, her nail polish, her quiet, the kind of music she loved.

Mornings and evenings, she lined up with her unit to receive her medications, an intersection of antipsychotics, a mood stabilizer or two, an antidepressant, a stimulant for attention, and a benzodiazepine, much like the blend she'd been on before the hospital. But the staff psychiatrist, who wore her blond hair in a ponytail and had the body of an aerobics instructor, was making improvements. She promised, with a perky, imperious confidence, that she would find the right combination of drugs and doses. Designing this cocktail was a central part of Caroline's treatment program at the hospital. At the sliding window, a nurse handed her two paper cups, one with water and one with her meds, and watched closely to be sure she swallowed the pills.

Seroquel seemed to be the psychiatrist's favorite antipsychotic. The cage was full of discussions about what it might do for you and what it would do to you. It became the featured ingredient in Caroline's cocktail. And the psychiatrist decided it would be best to wean her off her benzo, an antianxiety drug of the Valium family. This had to be done with care. When benzo doses are cut too rapidly, the brain can lurch into a chemical and electrical frenzy. The patient goes into a grand mal seizure. One afternoon, Caroline crumpled to the Hope floor, unconscious, her body in violent convulsions, her breathing in disarray.

She was belted to a gurney and was gone. She returned, days later, to a card from the Godfather and the other regulars of the smoking cage. "Come back," they had scrawled. "We miss you." "I'm really fucking scared." "We need our Black Hole." "I don't want you to die." "You mean so much to me."

She was part of them, smoking her American Spirits to the butt while joining in their games of Kill-Marry-Fuck. Each of them pondered aloud and picked one staffer to murder, one to wed, one to screw. This spiraled into long speculations about who the staff were when they weren't at the hospital, dispensing meds, policing group therapy, awarding privileges for compliant behavior and positive attitude—about who the staff were when they weren't concentrating on their paramount mission, preventing suicide. Whoever they were, Caroline felt that she was not and could never be anything like them, that even though she was close in age to some of the social workers, they, in their skirt suits, lived as if on another planet from her own. Her planet was the cage, where everyone joked about the regulation that patients not form friendships with each other, that they focus only on their own troubles and their own healing, a rule that was beyond ridiculous, everyone agreed, because what was more healing than relationships?

Her family visited for a requisite session. Her father told her that she was in the right place, that she should work hard in the hospital, do all the right things. "Listen to the professionals," he said. Her sisters talked about never knowing when or where they would find her blood, never knowing how to handle their fear, never knowing how to accept that their parents had nothing left for them. She brought her shame back to the smokers, put what she could into a few words.

What little she shared with anyone about her voices, how they sounded, what they said, she shared in the cage. She didn't confide much, but it was much more than she revealed to the psychiatrist or anyone else on staff. In response, some of the smokers commiserated, alluding to voices of their own. Then they all returned to their amusement: over the hospital's obsession with keeping them from killing themselves, over its bizarre regulations and changing prohibitions, like taking away her eyelash brush and minuscule bottle of mascara, over the staff's robotic devotion to dialectic behavior therapy, over how this, here, the conversation in the cage, was ten times more therapeutic than anything else that was happening, over another game: How would you kill yourself on this unit? They could raid the snack room, eat all the bagels, drink gallons of water, and cause the bagels to swell fatally in their stomachs. They could twist and wedge their heads through the cage bars, strangling themselves. They could sit on each other to the point of asphyxiation. They could take a toy Caroline had won in a hospital contest, an Answer-Me Eight-Ball Buddha that told your future, and drink the bit of blue fluid in the ball, liquid that the staff seemed to have overlooked and that would turn out to be lethal poison.

At the holidays, because she was Jewish, a patient asked if she wanted to light the unit's plastic menorah on the first night of Chanukah. Fire wasn't allowed; the candles were bulbs. Lighting meant turning them so they were screwed in. She stepped up but didn't know the blessing. Someone in a white metal back brace, who'd thrown herself off a bridge, spoke the prayer: *"Baruch atah, Adonai Eloheinu, Melech ha'olam . . ."* Caroline had a sense of it in English: "Blessed are You, Adonai our God . . . who has kept us alive . . . and brought us to this season." She stared down at her shaking hands, the words sucking her into a vortex.

Back in the cage, a woman, liver-spotted, turned abruptly. Her hair was plastered to the sides of her face. She drew on her cigarette and squinted at Caroline. "Oh, honey," she said, "you're going to be okay." But that wasn't the opinion of the staff. She was reticent, resistant, undermining. Hard to reach in therapy, her parents were told, and not reached enough by the adjustments in medication. If anything, her parents heard, months into her stay, she was regressing. There was no sound reason to keep her. It wouldn't be good practice. It wouldn't be serving anyone. The staff were sorry.

But, they said, she might well die if her parents took her home. The consensus was to send her to a therapeutic farm. Caroline wasn't included in the consensus, but she was given a choice of farms. She watched the promotional videos with the patients from the cage. The videos talked about "harvesting hope" and showed rows of vegetables and rough wooden fences and lots of cows.

"Are you going to wear overalls now?"

Patients made vats of cheese.

"What does your punk ass know about living on a farm?"

A line of baby pigs suckled at bulging udders.

Caroline said that besides her cat, Goo, the only animal she'd ever taken care of was Mabel, a pet tarantula who'd been murdered by an ex-boyfriend.

Sheep were shorn.

"How is our Black Hole supposed to take care of their black sheep?"

In one of the videos, cows traipsed across a snowy hillside and past a frozen pond; that place looked much too cold. Another farm was in the Midwest, much too close to Indiana. She chose the one in North Carolina, in Appalachian hills. It seemed like the farm of least suffering.

FIVE

David's wife, Amanda, drove him to his appointment with Dune. They crossed a corner of their city, passed the university where she taught screenwriting, and were surrounded by high-plains grazing land on the way to the lake where Dune had his home and office. There, as a mental and spiritual guide, as a medium, as a shaman, he administered his treatments. David was in disbelief that he was about to entrust his brain to a man with these self-assigned job titles.

Yet Amanda was in favor of this course and scornful of his vacillations. And his fear about what awaited him was overridden by so many others. The 2020 presidential election was near, and his depression, his searing nerve pain, his insomnia, his inability to reason or write or argue, and his inability to care about the issues that had defined his career were about to reveal him as a fraud and leave him jobless. Trump's 2016 election had bludgeoned him. The bullying, the lying, the immensity of the arrogance, the bluntness of the racism, the devastation Trump began to inflict within one week of his inauguration, with his Muslim ban—David had felt these as if they were physical blows to his head and body and had been grateful to work with an organization that could do battle against this man whom half the country wanted as its president. But now, with Trump threatening to subvert the vote as his reelection campaign faltered, David wished to be far from the litigation that was taking shape to counter Trump's chaos.

One issue that was likely to be fought over in court was the right to protest. If Trump lost and refused to accept the results, if he declared victory by insisting that only votes cast in person should be counted and all mail-in ballots should be thrown away, the Left would march in the streets. In the scenarios David's organization was planning for, Trump would probably send in the military or ICE or other federal agencies to squash the rallies. The organization divided its work into regions, and David's included states where the vote was expected to be close and the process of counting long, and where Trump's enforcement tactics could quickly get violent. The organization wanted to be ready to file emergency motions and even, ahead of that, to fend off Trump's federal troops with preemptive letters full of airtight arguments that Trump had no right to dispatch federal forces unless the states requested them.

The thicket of legal guidelines consisted of the Insurrection Act of 1807, the Posse Comitatus Act of 1878, and the language of the constitution itself, and these days David's brain felt like cotton, and he was having a hard time with something that normally would have been second nature: keeping straight which document provided the logic most advantageous to thwarting Trump and protecting protest. Another legal angle should have been even easier for him: the First Amendment right to freedom of expression. This was as straightforward as it gets in civil rights law. The organization had just invoked it when, in his state, local authorities had tried to shut down marches after the killing of George

Floyd by police. And the First Amendment was David's specialty. One of his First Amendment cases had gone all the way to the Supreme Court, and he'd done the oral argument in front of the nine justices, a crowning achievement for a civil rights attorney. But even the First Amendment's rationale now felt elusive, mired in fog.

In his depression, which left him, he well knew, utterly self-involved, his main worry about the upcoming election was not that Trump would again defy the pollsters and pull off a legitimate victory, nor that Trump would incite anarchy, destroy democracy, and retain power illegally; it was that David's own incompetence would be found out. He recognized how pathetic this was, how reprehensible it was, but the presidential election had become about him.

This unforgivable narcissism, too, would be discovered. Hadn't he shown hints of it three and a half years ago? As soon as the Muslim ban was decreed, his colleagues sped to the airport. People were being detained in back rooms and told they would be sent back to their home countries or to nations where they'd never lived. Families were calling, frantic. The lawyers needed to do interviews and compile information, to get temporary restraining orders. It didn't matter if their expertise wasn't in immigration law. They went. David didn't. He'd felt weirdly paralyzed. Something similar had happened weeks earlier, in the aftermath of Trump's mind-boggling win. David was standing at the edge of a basketball court, watching his daughter play, and was gripped by the need to flee. To where was unclear. Just elsewhere. Just out. Toward disappearance. Toward wherever he could forget the powerlessness Trump seemed to have injected into his bloodstream. Once, he'd been a fighter. These days, thinking about those moments of paralysis and wished-for flight, he saw, inside his impotence, an excess of self-protectiveness, an overwhelming panic on his own behalf.

As they drove through the scrubby grazing land and past scattered dormant oil rigs, he didn't speak about any of this with Amanda. She knew some of it, too much of it, already. She had been drawn to a man of purpose, a crusader, fallen in love with a man engaged with the world. Who was she stuck with now? How much longer would she stay? She had made a unilateral decision and was about to go on a three-week vacation by herself; the Airbnb was booked. Was she planning on returning, other than to settle on a system to parent their daughter? Amanda was an attractive, successful woman; her scripts, for TV and an independent film, had made it to the screen. Lately, she'd been sympathetic and even stalwart toward him, but she didn't always refrain from telling him how she viewed their marriage.

In the car, he clamped down on himself to avoid confessional mode. He was tempted to say something about their daughter, Gillian, that with her having just entered high school, he couldn't help thinking ahead to when she would leave for college and they would be without her. No more basketball games to attend. No more helping her to practice in the driveway. No more watching her favorite women's college team with her on TV. No more making Szechuan tofu stir-fry with her for Sunday dinners. No more family games of Scrabble. This loss seemed safe to express; Amanda would feel it, too. But it was four years away, and Amanda might hear his raising it as more evidence of the kind of man she was shackled with.

Sex hadn't happened between them for five months. This was by her count. The counts he was more conscious of were the months, accumulating toward two years, since he'd stopped his antidepressant and antianxiety medications. More self-involvement. Yet he wasn't incapable of sex or divorced from desire for his wife, and he had a frail sense that her occasional reminders of how long it had been were better than if she'd made no remarks at all. Still, sexlessness was like a solid wall he couldn't break through or find a way around.

He gave her foot massages with a device but could no longer take things beyond that. She said that she wanted a man who initiated. He no longer knew how and wasn't sure he ever had known.

She wasn't cruel; she praised him for the fight he was putting up against his symptoms; she'd never liked that he was on the drugs and never suggested that he go back on them; she told him how much she admired his swimming, his efforts at the ukulele, his refusal to give in. Last weekend, they'd hiked along a canyon rim, through arid shades of beige and rust, with cascading water eighty feet below. In this dramatic terrain, she asked what fantasies he had for the rest of the year. He understood afterward that he could have said anything, from "I want to have sex with you" to "I want Gillian to make the varsity team in her freshman year" to "I want Trump to lose in an unprecedented landslide." But the words that clogged his mind were "I want to be fucking undepressed," and he spared her those and marshaled no reply.

"You are so nonreactive," she said, dismayed that neither the beauty of this place nor her question reached him. "Do I have a partner? Do I have a living partner?"

He stared at a cluster of scraggly trees seizing the canyon wall and making do with whatever moisture existed in the rocky, red-tinted soil, and while he knew that people flocked to this landscape, he longed for the soft contours and low elevations and effortless, unthinking green of the mountains in the Northeast, where he'd grown up.

Now they drove by a Navajo reservation and mile after mile of strip malls. They found Dune's modest house a few streets back from the lake. Amanda left him there with a kiss and words of encouragement, and he followed the shaman down a flight of stairs and into a room that didn't seem shamanesque. There were no totems in sight. A pinball machine was the most prominent object. The lighting consisted of strings of white Christmas lights wrapped around a potted tree. But there were rugs in purple tones and a deep couch with lush blankets, and Dune himself matched his job description: long hair in braids, a dark beard, a white smock, a turquoise wristband, flowing pants.

David had lectured himself, in advance, about being judgmental. It was just a way of being resistive. What had his elite law degree, his sought-after judicial clerkships, his enviable legal career gotten him? He gave himself a swift review lecture as he and Dune sat down, he on the couch and Dune on the floor. Dune might look like a fortune teller in a rom-com, and yes, he was making money off what might be a New Age scam, but he seemed to sincerely believe he was helping people—and what was David going to get out of this if he didn't trust?

Trusting didn't turn out to be difficult. All Dune had to say was "Tell me something about why you're here," and David unspooled, answering, "I'm in a dark place. I mean, a really terribly, terribly dark place. Before this, before two years ago, I was depressed, I was anxious. There were reasons I took medication. But this, the withdrawal—at least, I think it's the withdrawal, because it started when I stopped the drugs—this is depression and anxiety on steroids. I was never what you would call an upbeat person, but I had a feeling of the world being okay. True, the world we live in right now isn't okay; it's been a shithole for the past four years; but think about the people in England in the middle of World War Two, in the middle of the Blitz, they still felt the breeze, the sun on their face. I want to feel the breeze and sun on my face. I used to come home from work and up to the front door, and next to the door there are windows, and I would see my wife and daughter hanging out in the kitchen, and I would think, *I love these people.* My parents didn't have the happiest marriage and our family didn't have the happiest household, and looking through the windows I would say to myself, I wanted this in my life and I have it. But that feeling in my heart is gone. I look in at my wife and daughter now, and they are people to cling on to, not people to love."

Dune asked that he recount a moment when he'd felt absolutely connected, to anyone, anything.

David let his mind wander, then said, "When my daughter was lying there—when she'd just been born—and I was about to cut the umbilical cord. I could feel my heart; I was like the Grinch when his heart expands three sizes. I was crying. I was out of control. My capacity for emotion was expanding and there was nothing I could do about it."

"And if you had to explain the depression before the depression, the sadness before the withdrawal, before the medication, where does your mind go?"

Slowly an image came. He described being young, being with his mother in their kitchen. She, much later, in her last years, became a heavy drinker, an elderly alcoholic, as if she'd finally discovered something that could deliver her. But back then, when he was a child and she was a woman on her way to two divorces, a woman whose hugs were rigid and whose approval was like a brittle prize, she had been making something in the kitchen, or maybe she had been cleaning up, and a glass container, the glass vessel of a blender, slipped out of her hands and splintered on the floor. She wept, but that wasn't the core of the memory. The core was his understanding, at whatever age he'd been, that she'd been upset beforehand, that the blender had nothing to do with her tears, and that he'd better be careful around her or she, too, would splinter. "I think I learned this loyalty to her."

"I find this very true."

"And I wonder if, for me, being anything besides unhappy would be disloyalty."

Dune said something about David's mother having advanced to a new phase of existence and that David might need to become acquainted with her in a new way. Dune said, "There is a state called joyful knowing." In David's hands, he placed a mug of warm liquid, the drink, as the shaman had already explained, that contained the psilocybin. It tasted faintly like mushrooms and more like weak chocolate.

DURING THEIR PRE-PSYCHEDELIC talk, they also touched on David's history with psychiatric drugs, picking up on what David had said when they first spoke by phone.

A therapist David had seen for years after college hadn't thought he needed medication. He didn't view David's depressive symptoms as amounting to anything that qualified as a major disorder. David felt they were working well together, and he had just experienced a strange sign of a breakthrough. It had occurred while he was sitting on the subway. The sensation was of an ice floe cracking apart in his head, the floe going instantaneously from solid to evanescent and clearing the way for unknown possibilities. There had been a moment of sheer wonder before he searched for explanations. He credited the therapist and their conversations.

At the time, he was putting all use of his Ivy League undergraduate degree on indefinite hold. His goal was to make the national crew team and compete at the World Rowing Championships. His college classmates were rising in corporations or finishing law school. The rowers he'd raced with or against were at business school or on Wall Street; they'd moved on. He had turned down a job that announced ambition and was working as a part-time crew coach and legal proofreader, so he could train. His parents were baffled and distressed. But this path made sense to him. He'd stumbled on his talent in high school and by college was one of the best strokes in one of the best boats in the country, yet year after year, in culminating race after culminating race, his boat had failed, coming up barely short in its final sprint, or thoroughly disintegrating, as if hexed, and falling so far behind that crossing the finish line felt like an empty existential exercise.

Some of these failures, in his judgment, had been his own isolated fault. Since his days on his high school team, he'd had a
habit of starting a race too strong; it seemed his muscles would not
obey his mind. And competitive rowing is a sport that uses energy
at extraordinary rates, bathing muscle cells in acid and stoking
excruciating pain. The body tries, self-protectively, to shut down.
You have to pace yourself. If you don't, if that level of pain comes
too early in a race, the oar in your hands becomes as useless as
a paddle. No one else in your boat may notice it, but to you it's
unmistakable.

This was David's story: if he'd just kept himself under control,
conquered his nerves and desire, those trophies, at those climactic
springtime races, would have belonged to him and his boatmates.
It was a story that fit into another he told himself, a wider and
longer one: he'd always been just outside, just short of, just separated from something—pleasure, contentment, delight, he wasn't
sure how to define it—that others possessed. In his senior year of
high school, he'd been nominated and elected class president; he'd
never been exactly an outsider. But once each year, by tradition,
on an undesignated day, the news came at mid-morning assembly
that classes were canceled, and the students cried out in exuberance and hurried to turn the building into a fantasia of concert
halls with makeshift instruments and fairgrounds with goofy rides
and venues for invented sports. His exuberance was fabricated,
his participation tentative, unnatural. He drifted from room to
room. All his classmates seemed capable of leaping into the day's
delirious state.

He harbored an assortment of memories like this, palpable
recollections running back to the age of five or six, emblems of
something he couldn't put a name to. But he believed that if he
could train hard enough, and race well enough at the tryouts to
put the near-misses of his college career behind him and earn a

place on the national team, if he could grasp this prize, if he could cross this boundary between outside and in, he would change something crucial about himself.

The tryouts, at the end of what turned out to be several years of training, went better than he'd let himself imagine. He rowed to easy victories in the early rounds. Rivals slapped him on the shoulder and predicted that he would be chosen—until he lost a tight sprint. In the next race, he was beaten badly. All had unraveled. The national coach told him that it was time to head home. It was time, as well, David knew, to plan out a life.

And he thought that maybe he should try something else. He raised medication with his therapist, who didn't object in principle but was mildly opposed in David's case. He assured David that he observed, in him, undeniable progress, reminded him of his experience on the subway, noted signs of a greater capacity for satisfaction, for joy. But the promise of antidepressants was in the air. Prozac, the first selective serotonin reuptake inhibitor, was fresh in public conversation. David asked repeatedly if medication could be a cure for whatever it was that plagued him.

The therapist referred him to a psychiatrist, and after a perfunctory evaluation, he walked away with a prescription not for Prozac but for a more time-tested type of antidepressant, a tricyclic. The pill did nothing for his sense of separation, his hovering though never crushing lack of happiness. Its impact was only in side effects: dizziness and head rushes from low blood pressure; desiccation of his tongue and the roof of his mouth. He stopped taking the medication after a year.

At law school, with his supplies of stamina and intelligence, he thrived academically. And he overcame a long-standing doubt about his ability to sustain romantic relationships. He settled in with an athlete he'd met in college. Eventually the relationship eroded, but his career was starting to flourish. He proceeded from clerkships to public interest litigation, and soon a beautiful screenwriter fell in love with a tall, impassioned champion for civil rights. They had a child. He sobbed—in ineffable connection— over the cutting of the umbilical cord. He could hardly see where to place the scissors.

Still, often, there was the lurking absence, the implacable idea that he was cut off from something others could easily claim. It was like being locked out of a room with a glass door, or staring at a glass case where everyone else could reach in and take some small, important token. It didn't matter that he knew that most lives were more difficult than his own; it didn't matter that he knew the extreme suffering of many of his clients.

He tried Prozac. This buffered his feeling of something amiss, allowed it to hang about him less, but even while it lifted some- thing away it imposed a weight, a fatigue. It flattened; it deadened. And then it snatched orgasm out of reach. He could not climax. It was torment. He could get partway there. Sometimes he could get close, could feel the faint beginnings. It was an ongoing mockery, a ritual of despair.

He visited a new psychiatrist, related his poor history with the tricyclic and Prozac, and listened as the psychiatrist recom- mended a third type of antidepressant, a variant on the second, a serotonin and norepinephrine reuptake inhibitor. There was no discussion about whether or not a third attempt was the right approach, about whether David's hazy symptoms of depression might be closer, on a spectrum, to the average state of mind than to a psychiatric crisis, no searching talk of definitions and diagno- ses. The assumption was: he'd been on medication, he should be on medication. It was, by this time, David's assumption as much as the doctor's. The therapist who'd thought differently was two thousand miles away and no longer a voice in his decisions. David filled a prescription for Effexor.

It spared him the sexual side effect and seemed—perhaps; it was hard to be certain—to keep the loitering sadness more at bay, letting it float near less frequently. But as the years passed, the drug never became an automatic, unconsidered part of his life. The pills, which he took each morning, posed repeated questions: Who are you? Who are you without putting this substance into your system? Who are you underneath?

So when Trump's election filled him with the urge to flee and saturated him with feelings of powerlessness and paralysis, and when Amanda, watching him struggle and listening to his self-castigation, said that the medication didn't seem to be doing him much good, if it was doing him any good at all, he decided to think about, at least *think* about, tapering off. He was not one to rush into anything. He consulted a psychiatrist who advised that a six-week taper would be gradual enough. He read online and resolved on three months. He took his last fraction of a pill on a morning in late springtime, 2018, and had the sense, within weeks, of a toxic sludge invading his extremities. He found himself tugging at his fingers and wringing his hands. As for his depression, what had never been worse than a hovering and intermittent discontent became an unbearable oppression, an occupation of his entire body.

And he could not sleep. He'd never had trouble with insomnia. Suddenly he was waking at one or two in the morning and could not find his way back into unconsciousness. Night after night, week after week, this persisted. He expected to become so enervated that he would stay asleep for seven or six—or he would have been thankful for five—hours, but mounting exhaustion made no difference. His mind was decomposing. To stave off an outright emergency, he returned to the psychiatrist who'd counseled him about weaning off the Effexor, and was given a prescription for a sleeping pill, temazepam, a benzodiazepine. If he was cautioned against using it chronically, he didn't hear the warning. He took a pill every night; he slept not well but not quite so minimally, and the psychiatrist kept him supplied.

Eleven weeks later, Amanda pointed out that benzodiazepines are addictive, that her Google searches produced lines like "should be taken for short periods only (for example, two to four weeks)" and "the use of benzodiazepines may lead to dependence." He titrated himself off the drug carefully, taking five months to do it. The scorching greeted him predawn, during and after the tapering. His feet, ankles, hands, just beneath the surface of the skin, were aflame. His neck. Across his shoulder blades. His lips burned. His belly, face, ears. He stood under cold showers. He covered the worst spots with ice packs. But there was nothing to be done except wait out the subcutaneous pain that was so consuming it seemed to illuminate the ending of every agitated nerve.

He breathed. He waited. When the burning faded, he scoured the internet for information. He read, "My whole body is burning and sizzling like I am constantly being electrocuted." There was a kind of netherworld of websites devoted to withdrawal from antidepressants or benzodiazepines or both. "Had to stop all the clocks in the house because their ticking sounded unbearably loud." "My shoulder blades, upper arms, and legs are burning like hell." "Cog fog." "Tinnitus."

He descended into testimonials. He studied explanations, more clarifying for the chemistry of benzo withdrawal than for coming off antidepressants. "All benzodiazepines act by enhancing the actions of a natural brain chemical, GABA (gamma-aminobutyric acid)," he learned on the site of a psychopharmacologist. "GABA is a neurotransmitter, an agent which transmits messages from one brain cell (neuron) to another. The message that GABA transmits is an inhibitory one: It tells the neurons that it contacts to slow down or stop firing." He read on, "GABA has a general quieting influence on the brain: It is in some ways the body's natural hypnotic and tranquilizer. This natural action of GABA is augmented by benzodiazepines, which thus exert an extra (often excessive) inhibitory influence on neurons." Remove the drug from a brain that has surrendered to it, and counterbalancing systems, long suppressed, were unbound. "Nearly all the excitatory mechanisms in the nervous system go into overdrive." He went back to the testimonials: "Electric shocks." "Skin and scalp so sensitive it feels as if insects are crawling over them."

More and more months crept by. Every aspect of his agony remained undiminished. His only consolation was the alliances he struck up, by email or text, with people he would never otherwise have known in places he knew nothing about. He marveled at the improbability of these friendships. There was a woman mired in withdrawal in West Virginia. There was a man with a hopeful story in Iowa.

David tried to parse causes. The nerve pain he could link to cutting off his benzodiazepine. The insomnia could be traced to quitting the Effexor. The rest of his symptoms—his despondency, his dysfunctional intellect, the replacement of love with abject need when he looked at his wife and daughter—how these had arisen he could not know, except to be sure, or as sure as he could be of anything given the state of his brain, that they were tied up with the medications and then the medications' absence within him.

TWO FAILED DRUGS, one for tuberculosis, the other for schizophrenia, led to the antidepressants David had tried. In clinical trials, in the 1950s, the TB drug didn't do well in fighting the bacterial disease, but patients who took it felt euphoric. At one TB sanitorium, bedridden women, their lungs ravaged, began dancing on their ward. Nathan Kline, the celebrated, Lasker-winning psychiatrist who promoted "massive doses" of Thorazine as the optimal treatment for psychosis, took the TB drug himself in a three-month personal experiment. On it, he needed only three hours of sleep per night and "woke up feeling fine," he reported to *Newsweek*. "I just can't believe that God made the human machine so inefficient that it has to shut down or be recharged one third of its lifespan." The medication seemed to have potential as an antidepressant—but once it was on the market, a pattern emerged. It assaulted the liver, caused high rates of hepatitis, and left dozens of patients dead.

Meanwhile, though, the failed schizophrenia drug also sparked an unexpected brightening in mood, and together the two drugs gave rise not only to the first class of modern antidepressants, the tricyclics, named for the trio of carbon rings in their chemical structure, but to a theory about how the drugs provided relief, a theory applied to all antidepressants since—a theory that almost anyone who has taken or considered taking an antidepressant, or anyone who has spent a bit of time thinking about the psyche, is familiar with to this day.

The theory was built on experiments with rabbits. By injecting a sedative, researchers made the animals sluggish and apathetic in ways that seemed to parallel human depression. Then they found that they could protect the rabbits against this depressive effect by pretreating them with the failed tuberculosis or schizophrenia drugs. And they linked the rabbits' depression or protection to levels of serotonin. In the early '60s, Julius Axelrod, who would win a Nobel for his work, shed light on what was happening by showing that neurotransmitters, after creating the chemical bridges that permit signals to cross the synapses between neurons, are reabsorbed into the presynaptic cell, where they are saved, inactive, for later use. One of the two drugs blocked this reuptake. It left neurotransmitters actively bridging the gaps, keeping neurons connected, keeping signals robust. The other drug blocked a different process of draining neurotransmitters from the clefts, a process of breaking them down and flushing them away.

By the mid-'60s, all of these discoveries, from the dancing TB patients to the rabbits to Axelrod's findings about neurotransmitter reuptake, fed a hypothesis put forward in a paper from the NIMH, proposing that "some, if not all, depressions are associated with an absolute or relative deficiency" of neurotransmitters in the brain. The author of the paper, Joseph Shildkraut, who would go on to write about the intersection of depression and spirituality in art, was wary of his own reductionism; he emphasized that his hypothesis was "at best a reductionist oversimplification of a very complex biological state." Nevertheless, scientists and pharmaceutical companies and, in the end, everyday practitioners and the public, latched on to what became known as the chemical imbalance theory. For a while, because Alexrod focused on norepinephrine, a dearth of this neurotransmitter was seen as the reason for depression. But ultimately, depression was blamed mostly on deficient serotonin.

The language of chemical imbalance was soon applied not only to depression but also to psychosis, with the condition attributed to excess dopamine instead of insufficient serotonin. The concept of imbalance was essential to the triumph of the biological within psychiatry around 1980, and from the late '80s onward, with the introduction of Prozac and its competitors, the theory became a theme in appeals to doctors and patients to prescribe and take antidepressants. Loosened federal laws, starting in the mid-'80s, let companies advertise prescription drugs directly to consumers, something only one other developed nation, New Zealand, allows. Eli Lilly promised that Prozac would "make you feel more like yourself by treating the imbalance that causes depression." One holiday-season ad from the '90s displayed a spindly, Charlie Brown–style Christmas tree, wilting under the weight of a single ornament, side-by-side with a full-bodied, sturdy evergreen topped by a radiant light. Prozac, the text read, would "bring serotonin levels closer to normal." It would cause only minor and fleeting side effects, such as "upset stomach and headaches." The drug was right for anyone with feelings of sadness or loss of energy that "last more than a couple of weeks." The ad concluded: "Prozac has been prescribed for more than seventeen million Americans. Chances are someone you know is feeling merry again because of it."

The manufacturer of Paxil, GlaxoSmithKline—the corporate descendant of Smith, Kline, and French, distributor of Thorazine—put things this way on its website in the 2000s: "Just as a cake recipe requires you to use flour, sugar, and baking powder in the right amounts, your brain needs a fine chemical balance in order to perform at its best. Normally, a chemical neurotransmitter in your brain, called serotonin, helps send messages from one brain cell to another. This is how the cells in your brain communicate. Serotonin works to keep the messages moving smoothly. . . . Paxil helps maintain a balance of serotonin levels. . . . Paxil is with you throughout the day to help you manage and treat your condition." Pfizer TV commercials explained the same chemical problem and pledged that "prescription Zoloft works to correct this imbalance. You shouldn't have to feel this way anymore." Pfizer had a variation for Effexor; the pill would help symptoms subside by "correcting an imbalance" in norepinephrine as well as serotonin.

Imbalance, with its suggestion of disproportion, has always been something of a misnomer. It sounds scientifically precise, but no one has charted ratios in the chemical ecosystem of the brain. The concept actually boils down to a shortage, and scientists have tried to link depression to subpar levels of serotonin. In 1969, a Yale University psychiatrist found evidence of only slightly lowered serotonin in the cerebrospinal fluid of depressed patients. In 1971, researchers at McGill University discerned no statistically significant difference in serotonin measures between depressed patients and controls. In 1974, the Yale psychiatrist conducted a follow-up to his earlier work and registered normal readings of serotonin in a new set of depressed subjects. Data collected by a Swedish psychiatrist led her to argue, in a 1975 paper, that deficient serotonin did indeed lie behind depression—but a reanalysis of the study showed that plenty of its nondepressed subjects fell into a category with the lowest levels of the neurotransmitter. An NIMH study, published in 1984, determined: "Elevations or decrements in the functioning of serotonin systems per se are not likely to be associated with depression." Research done by a University of Texas psychiatrist demonstrated, in 1999, that depleting serotonin did not cause depression in healthy subjects.

None of this meant that serotonin had no role whatsoever in depression. The most generous data on the efficacy of selective serotonin reuptake inhibitors, the SSRIs, indicates that—if comparison with placebos is disregarded—around half of depressed patients benefit, to some extent, from the drugs. So targeting neurotransmitter reuptake seems to have an unknown relevance. Most likely, serotonin's part in depression is one of myriad supporting roles. Nestler liked to point out that, for those who do find some relief in pills that leave more serotonin in the synapses, the relief is delayed, that the drugs take a good deal of time to kick in, a sign that simply blocking reabsorption isn't enough. The blocking happens within minutes; benefit can take more than a month. The same is true with the serotonin and norepinephrine reuptake inhibitors, the SNRIs. This is almost surely because mood depends on a still unspecified array of interactions, of molecules and mechanisms, spread throughout the brain. Some of these may be indirectly—and not immediately—affected by serotonin or norepinephrine. Some may lie completely beyond the influence of these neurotransmitters.

Nestler and a number of his colleagues had been dissatisfied, since 1990, with theories of depression that zeroed in on neurotransmitters and synapses. He had long thought that better keys to mood would be found within, not between, the cells of the brain. Yet despite all the evidence that complicates or contradicts the notion of neurotransmitter deficiency, the explanation has persisted. A *Charlie Rose Show* roundtable of esteemed scientists, in 2009, used a graphic of serotonin floating in a synapse as a basis for exploring depression. Intricacy and the limits of current research did seep into the discussion, but toward the end of the hour, the Nobel laureate biochemist Eric Kandel proclaimed with cheerful authority, "This is like diabetes"—that is, the reuptake inhibitors could address depression as surely as insulin deals with spikes in blood sugar. Levels just needed adjusting. A 2021 online guide from the NIMH, written for laypeople looking for insight, tells readers that "even though not all details are known, experts believe that depression is caused by an imbalance of certain chemical messengers (neurotransmitters) like serotonin." Nowadays, uncertainty does get noted—before scientific expertise is asserted.

But when the effect of placebos is taken into account, thorough uncertainty descends around the science. The word "placebo" derives from a Latin phrase meaning "I shall please." In medieval France, people who showed up at the funerals of strangers, ready to sing for free food, were called placebo singers. Serving in the American military in World War II, Henry Beecher, an anesthesiologist, ran out of morphine, infused wounded patients with saline solution, and told the soldiers he was giving them a painkiller. Nearly half said their pain was eased. A paper Beecher published, in 1955, led the FDA to insist on placebo-controlled trials as part of the approval process for most new drugs. And beginning in 1998, Irving Kirsch, who is the associate director of the Center for Placebo Studies at Harvard, published a set of papers and, in 2009, a book, *The Emperor's New Drugs*, arguing that the most popular antidepressants were barely better than placebos at battling depression—and were probably no better at all. When the medications were seen for what they were, he maintained, it was clear that they drew most or all of what power they had from the placebo effect.

By filing freedom of information claims with the FDA, Kirsch had collected all the data that had been sent by drug-makers to the agency as approval was sought for Prozac, Paxil, Zoloft, Effexor, and two other top-selling antidepressants. To win the agency's okay, companies are asked to submit the outcomes of all their trials, yet the FDA's standard for approval is that just two trials need to show efficacy and safety, no matter how many failed trials have been run in order to get the two passing grades. This meant that Kirsch gathered a wealth of results that had never been published, since the companies had no interest in airing their negative outcomes.

Putting all the studies together, the passes and the fails, and conducting what is known as a meta-analysis, Kirsch exposed what the companies had already confronted and concealed: the degree to which placebo pills rivaled the medications at treating depression. Placebos were eighty-two percent as effective as the drugs. Said another way, eighty-two percent of the impact of the medications was due not to ingesting a chemical that inhibited reuptake but to psychological factors involving self-persuasion and illusion. Hope, expectation, the attention of a doctor or nurse, the ritual of taking a pill—the illusion might be fed by a host of forces.

To make things more numerically vivid, Kirsch guided readers of his book through a system, the Hamilton Depression Rating Scale, widely used to assess severity of depression. A physician-administered questionnaire produces a score from zero to fifty-two. On this scale, which the FDA relies on, medications outperformed placebos by less than two points, a differential that did reach the threshold of statistical significance but that was clinically meaningless, signifying next to nothing in terms of how patients felt. "A two-point difference can be obtained by being less fidgety during the interview or by eating better," Kirsch wrote. He listed a smattering of improvements and continued: "Any one of these changes can make a two-point difference in a person's depression score, even if there are no changes at all in the person's depressed mood, feelings of guilt, suicidal thoughts, anxiety, agitation, or any of the other symptoms of depression."

Kirsch had spent much of four decades examining the medical role of placebos, and he allowed that the phenomenon he'd isolated in depression also appeared in studies of pain and other conditions tied to the psyche. Yet the magnitude of the placebo response in the antidepressant trials was striking. Then he highlighted something else that was crucial. People who sign up for antidepressant trials probably anticipate side effects. They've been looking for medical help; they've heard about side effects or know them firsthand if they've tried other drugs in the past. In a trial, some who are randomly assigned to the placebo group will guess this as soon as their pills cause no discomfort or disruption. Their suspicion will affect their hopes and dull the placebo's influence. The placebo effect was inherently undercalculated.

But what if a placebo with side effects was used? Not serious side effects, like impotence or anorgasmia or weight gain, but more benign ones. Nine of the trials Kirsch investigated had employed what are known as active placebos. They caused dry mouth and headaches, helping subjects to believe that they were swallowing the real thing. In seven of the nine trials, statistical significance disappeared altogether. Active placebos equaled the medication.

Dosage didn't matter. High doses of antidepressants didn't create an advantage over placebos in the FDA data. Deepening his picture of antidepressant mirage, Kirsch described research by a German psychopharmacologist: "Depressed patients who failed to respond to antidepressant medication were given an increased dose of the drug, following which seventy-two percent of them improved. . . . The catch was that the dose had only been increased for half of the subjects. The others only thought the dose had been increased; in fact, it had not. Yet the response rate was the same seventy-two percent in both groups."

Kirsch's research brought him a wave of media astonishment and admiration. "This is huge," Lesley Stahl said, interviewing him on *60 Minutes*. It also brought on a backlash. At the very least, his critics argued, antidepressants were a valuable treatment for those suffering the worst depressions—though this wasn't much of a critique, since Kirsch had said the same in his book, with a graph distinguishing the worst-off as gaining a minor but not negligible effect from the drugs. Peter Kramer, practicing psychiatrist and author of *Listening to Prozac*, spearheaded a broader attack with an essay in the *New York Times* and then a book of his own, driven by outrage over Kirsch's work and defending the use of antidepressants for mild and moderate cases. The trials Kirsch had analyzed were flawed in their design and execution, Kramer insisted, skipping past the fact that the companies themselves had designed and executed them, so that any bias was likely to run in the drugs' favor.

Yet among Kramer's main issues was that Kirsch's conclusions were statistical rather than personal. From his own experience in treating the mildly and moderately afflicted, Kramer asserted his "clinical wisdom" about the drugs' power. He'd seen it; he'd witnessed it; he knew it. He dismissed the placebo data. He argued for the chemical effect of antidepressants, but, paradoxically, his book frequently argued by—and for—anecdote. "Doctors don't see averages," he wrote. "They see patients." He didn't mention any misgivings about this unscientific logic; he didn't seem aware of the hubris within it.

Kramer's side appeared to win out. Or perhaps this had less to do with the criticisms of Kirsch than with a widespread need to believe in the drugs. Antidepressant use continued to soar, rising by a third among American adults between 2009 and 2018. Kirsch seemed to be a man on his way to being forgotten.

Then I talked with Marc Stone, a deputy director at the FDA's division of psychiatric products. From within the agency, he had just expanded on Kirsch's explorations, analyzing a wider set of trials, some completed as recently as 2016, covering twenty-two approved antidepressants. Stone had, in all, over two hundred and thirty studies with over seventy-three thousand subjects, and unlike Kirsch, who had worked with each trial's overall outcomes, Stone had each individual subject's data, so he could run a closer analysis. On the differential between drugs and placebos on the depression scale, his result was a hair below Kirsch's: 1.75 points. Not nothing, but near to it. On a graph, the drug and placebo lines tracked each other tightly. "They overlap substantially," he said.

When I checked in with Stone again, he and his co-authors were on the brink of publication in the *British Medical Journal*. And they'd run a more sophisticated analysis. A distinct subgroup of subjects—around fifteen percent—were helped, sometimes a lot, by the medications. The rest got no meaningful benefit beyond the placebo effect. The problem was that there was no way to reliably characterize who was in the fifteen percent. Severity of depression correlated only weakly. The new results were a caution that if you didn't happen to have what Stone described as the undetermined neurological on-off switch that was reached by the drugs, then you were out of luck, and were risking side effects with no hope of advantage. He believed that the study should lead to a rethinking about how antidepressants were prescribed. "If a patient is very depressed," he said, "you might still want to prescribe—despite the low fifteen percent chance—because of the severity. But for those who are less depressed, you would want to be much more hesitant until we understand more about who will respond. These may not be the most toxic drugs, but they do have important adverse effects."

Stone didn't expect his research to be well received. "The doctor's experience is that the patient comes in, and you give them this drug, and some of them are much better in time, and you have your own reward system that says you made the right choice, that you were right to medicate and that you chose a drug that worked. There's cognitive dissonance; the strength of the placebo effect gets ignored. You tell yourself you did a good thing, that you're a good doctor. Then you have a study like this. There's going to be denial."

DAVID DRAINED THE cup of mushroomy chocolate, and Dune told him it could be fifteen minutes to an hour before he felt anything. The reason he was here escaped him. How was psilocybin supposed to help with his symptoms of withdrawal? How was that even plausible?

It had been easier for David to learn online about the aftermath of cutting off benzos than about stopping antidepressants; there seemed to be more science about the benzos. The mysteries surrounding the benefits of SSRIs and SNRIs, all the unknowns about the changes their reuptake blocking might generate throughout the brain, were multiplied when it came to the changes of tapering off. The science was nonexistent. But antidepressant withdrawal was just starting to get attention as David was going through it, and from stories in the *New Yorker* and the *New York Times*, he knew his symptoms weren't imagined—or almost knew, because self-doubt taunted perpetually.

"'I couldn't finish my college degree,'" he read. A woman recalled the fatigue and confusion that came with weaning herself off. The stories described vertigo, surges of anxiety, insomnia, emotional numbing, paranoia, sensations of electrical zapping shooting through the brain. A woman was beset by an oversensitivity to color; the bright hues of a cereal box were too much. A psychiatrist confessed his skepticism about the reality of withdrawal symptoms until he took himself off his own antidepressant. He'd only been on the drug eighteen months, and had spread the tapering over a year, yet even so he was ambushed. David read about "dystalgia"—a feeling of all-consuming futility.

The problem had long been overlooked, partly because psychiatrists tend not to follow patients closely over extended periods, prescribing instead based on quick consultations, and partly, David learned, because the pharmaceutical companies were well-practiced in dismissing the issue of withdrawal. For decades, they'd discounted the addictive properties of Valium and the rest of the benzos, and before that they'd done the same with barbiturates, benzos' precursors. Opioid addiction was waved away in the 1990s and 2000s, while the industry persuaded doctors that patients were suffering from an epidemic of untreated pain and that oxycodone and its cousins were the needed remedy. One pharmaceutical corporation sought advice from the consultants of McKinsey & Company, who advised, according to a lawsuit brought by the state of Massachusetts, that opioids be promoted not only for addressing pain but for broad psychological effects like making patients "more optimistic and less isolated."

The physical dependence engendered by antidepressants doesn't come with anything like the dire risk that accompanies opioids, but the pattern of corporate denial and dismissiveness has been similar. Eli Lilly produced a report, in 1997, on what it termed "discontinuation syndrome" with antidepressants. The report arrived at three conclusions: the symptoms were "generally mild," Lilly's own drug, Prozac, was associated with fewer discontinuation problems than other antidepressants, and any difficulties could be "rapidly reversed by the reintroduction of the original medication."

David read that the numbers of U.S. patients who'd been on antidepressants for five years or more had doubled since 2010 and tripled since 2000 and that some of these people were caught in a kind of medical trap. Patients who tried to get off the drugs— some because of the sexual dysfunctions that themselves were downplayed by the pharmaceutical companies and ignored by practitioners but that affected over half of those who took the drugs, and some because they yearned to rediscover underlying selves and lost identities—were assaulted by symptoms but told by their doctors that what they were going through was just a return of their old depressions. A paper in *The Lancet Psychiatry* explained: "Although the withdrawal syndrome can be differentiated from recurrence of the underlying disorder, it might also be mistaken for recurrence, leading to long-term unnecessary medication."

This is not who I was, David sometimes shouted silently to himself. The tenuous reasoning of his turning to Dune was that the psilocybin would give him a "mystical-type experience," as the Hopkins professor's TEDMED talk said, that the cosmic would override the chemical wreckage, that new connections would be made between regions of his brain, that new connectedness would be created between him and the world, that a revived connection would take hold between him and his work, that his relationships with Amanda and Gillian would be restored to the kind of love he'd known with them—with the pleasure he'd once taken in rebounding for Gillian in the driveway and feeding her passes at the perimeter, or the pleasure he'd once taken in Amanda's spontaneity when she'd rented an RV and the two of them spent a weekend just staring at swamp turtles in a Louisiana bayou—rather than a love defined by fear that he could not survive without them and that they would both soon rescue themselves by severing their lives from his self-obsessed desperation. And the way Dune talked, anathema as it was to David's legally trained mind, seemed to say that all this was possible. David needed, he said, while they waited for the psilocybin to take effect, to have a conversation with his dead mother. Serene notes of New Age music played softly. She had, Dune assured him, "graduated from that past state of being"—rigid, brittle, breakable—"into the light. She is able to feel joy and fulfillment in the journey of her existence." David needed to choose to join her on that journey. "Allow yourself to have a conversation with her now."

"Oh God," David said, "is this the burning nerve pain?" There was a tingling. The psilocybin seemed to have triggered the scorching.

Dune told him no. The sensation faded. David tried to calm himself. He thought again that he was the type of person who wouldn't react well to the psychedelic, that this was not safe—or that he wouldn't react at all, that he was too rational, beyond its reach. The music became not louder but somehow nearer. He pitched downward into a climate of colors. Fragments of blue, yellow, pink enveloped him. The fragments formed shapes he couldn't name. They swirled and swept him into them. He closed his eyes, opened them, closed, opened—the iridescence remained. Dune shimmered. Someone, David himself or his guide, he couldn't sort out which, said, "Can you imagine carrying a child inside you but being unable . . ." Dune was himself and not. David, on his back, felt himself raise his hand, forefinger lifted, to make a point. The colors grew sharply geometrical. ". . . do you know how wounded she must have been for that to be the case?" There was a fuzziness around Dune's face, his shoulders. The geometries morphed into objects that were both plainly flowers and unidentifiably abstract. It was the same with the object that was both a fetus—there was its tiny head—and nothing but a blur. He said nothing about this to Dune. It morphed back into abstract flowers and again into something bordering on prenatal. Dune said, "It's important for you to be by yourself for a few minutes."

Someone spoke a chain of logic, its components unspecified: this then this then this.

Dune said, "I'm going to step out."

There's no need, David said but didn't say. *I don't want you to.*

SIX

Whurt hen the three of us stood in my brother's childhood bedroom, when our mother said, "Give him the book," and when our father left and returned and placed it in my hand, they were in search of an ally: someone who would learn the book's lessons and accept them as truths just as they had newly learned and accepted them. But learned and accepted are weak words for what they had undergone and what they wished for in me. It was as if they had ingested the book's sentences and elevated its paragraphs to articles of faith. They were immediate converts. They wanted a profession of faith from me. If I, their older and more conventional son, their steadier son, would join them in their belief, we would form something sturdy and stable, unified in un-questioned knowledge.

I doubt it occurred to them that I might have any reservations. Their faith, generated, like many faiths, out of need, was too swift and strong for questions. Their son considered himself blessed with otherworldly gifts. It wasn't only that he might be the messiah and that he could cure our grandfather's Alzheimer's. He was also connected with Joan of Arc—because he and I happened to have been born in a French town that was significant in her story—and he planned, after saving our grandfather in Brooklyn, to proceed to France, so he could be in proximity to the fifteenth-century prophet. He had a disease of the brain and required brain medication. Without it, a single prospect loomed. "Suicide . . . attempted suicide . . . suicidal feelings . . . suicide . . . hanged herself" reverberated through the book's opening pages. But with medication—just as with so many medicines of the body—his brain would be brought back to normal. He would be himself. There was nothing to question. Skepticism was beyond our parents' imagining. This was their son, my brother, who needed to be rescued. The one fortunate thing in this emergency was that there was a means of rescue. Our father handed me the book, and they waited for me to read enough of it to express my agreement and credence.

Our mother, to emphasize the book's urgent information, said that the book was new. This wasn't quite accurate. What I'm describing happened in 1983, and the book, *Moodswing*, by Ronald Fieve, a psychiatrist at a Manhattan hospital affiliated with Columbia University, had been published eight years earlier. Though it declaimed "chemical advances" that already amounted to a "revolution" throughout psychiatry, it was more a harbinger of the revolution that was on the way—and, too, of all the books with similar messages that would be written by authors with like-minded missions in the next decades. But I couldn't have foreseen any of this at the time. Except for a few psychology courses in college, classes taught by a behaviorist department most interested in training rodents with rewards and punishments, I was clueless about the field.

Yet more than the hyperbolic coloring of the book's covers and the hyperbolic sound of the back-cover pronouncement about "new medicines for the mind," something between the covers made me uneasy. Fieve did what many authors have done since;

he took the psychiatric diagnosis at the heart of his mission, bipolar disorder, with its swings between extraordinary highs and lows, between mania and despondency, and applied it retrospectively to historic figures. Theodore Roosevelt, he wrote, had been bipolar. The evidence consisted of Roosevelt's charisma, his need for little sleep, his eagerness to lead the cavalry charge up San Juan Hill, his "zealous, manic" campaign for the vice presidency on the McKinley ticket, during which "he seemed to be in the grip of a messianic urge, commonly seen in manics," and his habit, as president, of devouring Greek classics after midnight.

Fieve diagnosed Abraham Lincoln as bipolar, with more lows than highs, based on the hereditary evidence of his father being "dissatisfied and restless in his youth," and on Lincoln himself going through an "overtalkative, wild" phase in his adolescence and early twenties. "On one occasion," I read, "when he was not invited to a wedding, he wrote a longwinded, inappropriate, and insulting poem to the bride and groom." Lincoln was despondent after the death of his first love and later, as a state senator, suffered a bout of melancholy that "lasted for more than a week," kept him away from legislative sessions, and caused some of his colleagues to worry that he would commit suicide. Fieve said that proper treatment today would be to "insist on hospitalization, observation for suicidal intent, antidepressant drugs, and later administration of lithium as the treatment of choice for such a condition."

Ernest Hemingway was, for Fieve, defined by the disease of manic depression, another term for bipolar disorder. The symptoms were, on the one hand, Hemingway's "abundant energy," the fact that "when he was not writing, he was fighting, or deep-sea fishing, or hunting," and, on the other, his stretches of fierce self-criticism, his heavy drinking, what Hemingway called his "black-ass days" and "bumping off" periods. The final proof was his suicide.

Sitting on my brother's bed, I read for an hour or so before finding our parents in the kitchen. I understood how much solace they took in the book I now held. My brother not only had a circumscribed disease that was highly treatable; he was also among millions of Americans, Fieve said, who went undiagnosed in the present as well as grand figures who had gone undiagnosed in the past. Our parents awaited my statement of faith, but I was, quietly, in a state of revolt. I was twenty-three, yet even now, thinking back, I rebel. A psychiatrist was taking the singular lives of great human beings and reducing them to an illness. It felt—and feels—like an act of intellectual violence.

THIS WAS NOT merely what I felt that Fieve was doing; I sensed that our parents were doing it with whomever my brother was at twenty-one and whomever he was destined to be.

A year and a half ago, my downstairs neighbor, a spirited visionary, who opened a series of fledgling, closet-sized galleries that changed the art world in Toronto, and the founder of an art space in a Manhattan alleyway, killed herself. She left behind two teenaged sons. She wasn't living downstairs at the time. She had sublet the apartment to a younger friend; he was the one who told me. "She had a mental illness," he said, "and she wasn't getting treatment." The news of her death was devastating—I pictured her younger son playing like a straw-haired nymph in the flowery jungle of the back garden—but the words of her friend, almost as much as her death, left me speechless. This was what he had been taught to believe. This was how he found comfort. This was how he shut the door.

I'm not much for confessional writing; I hesitate to include the fact that extended periods of my adult life have not gone by without my thinking about suicide. I regret that double negative. It reflects my resistance to extracting this truth from myself and exposing it on the page. Though I have hinted at it, in passing, with one or two people in my life, I have laid it bare for no one.

But it is inescapably relevant here. Have periods of more than a year gone by without these vivid thoughts? Possibly, but I doubt it. Mostly, they are something of a constant, pushed away but forever returning, welcomed sometimes, invited in, asked to stay for contemplation and planning, then either banished in a hurry or asked politely to leave.

To a psychiatrist's inevitable next questions—Are my thoughts accompanied by a method or methods? Do I have a means in mind?—the answer would be yes, I do. And as I write this, I am overtaken by the impulse to fend off psychiatry's—and society's—reaction. I am rushing to say that I bet such thoughts are more ordinary than we care to believe. I am rushing to deflect alarm, to avoid reductive perceptions and protective boxes, to evade what comes with such a confession: the feeling of being alien, of failing to sufficiently value life, the sense of moral failure, of falling far short.

I will include one more personal detail, related but distinct: the sensation that my body, my upper body especially, my arms above all, are turning to water, that these physical parts of me, whose strength at other times gives me some satisfaction, are liquefying. The water is heavy. There is nothing flowing or graceful about it. It is water, but it is leaden.

Perhaps strangely, the thoughts of suicide are much more frequent, and usually less troubling, than the feelings of water. Suicide is a way out, there if I need it, calming in its availability, though also frightening because it threatens to assert a power of its own, beyond intention. The liquid is much less manageable.

Once, not long ago, I was invited to a small dinner. It was billed as a sort of literary salon, and the evening was built around me. I'm easily flattered by any attention I get, but this gathering included two or three psychiatrists, and the book of mine that we were discussing dealt with aberrant erotic desire and unusual, unsettling love, and through a sequence of restrained debates,

debates around what could and could not be legitimately designated as disease, things got testy. I turned to one of the psychiatrists, a man in private practice. He had just said that he rarely—very rarely—saw a patient for whom he didn't prescribe medication. He was not an automaton of profit-maximizing, fifteen-minute evaluations. He projected an aura of wisdom and had a position at an elite university. I asked if I could pose a sophomoric question based on what might be a false dichotomy, but a question that might be clarifying. Take Hemingway, I said, and take the common psychiatric judgment that he was manic-depressive. Now, let's say that you're faced with a choice—and I emphasized that some would argue it was a false one—between medicating his manic depression beginning when it most likely first appeared, sometime in his teens or twenties, and then continuing the medication as recommended for the rest of his life, and in the process eliminating *The Sun Also Rises* from literary existence, or, alternatively, leaving him to the unmedicated fate of eventually shooting himself at the age of sixty, but allowing for the creation of that great novel, other novels that were good if not great, and a number of American literature's best short stories. The psychiatrist didn't pause. Presented with this choice, he said in self-assured tones, he would medicate.

FINDING OUR PARENTS in the kitchen, I returned the book to our father. I handed it to him gently. I can only hope that whatever I said was gentle as well, but I know that I didn't keep my thoughts entirely to myself, because while I can't recall my words, I must have expressed something about my objections. I remember our father's face looking stunned and all the more drawn, and our mother, for a moment, lashing out. I hope that I apologized.

)

OVER THE PAST few years, as if I were still in my brother's annex room, sitting on his narrow bed, surrounded by the Aboriginal bark paintings, I have read many of the countless books of popular science and works of literature, published in the 1980s, '90s, and 2000s, that delivered the doctrine. Nancy Andreasen, a psychiatrist who was given the National Medal of Science by President Clinton, was most closely aligned with Fieve. In 1984, in *The Broken Brain*, she wrote about her discipline's move "from the study of the 'troubled mind' to the 'broken brain,'" and likened psychiatric disorders to "heart disease, muscular dystrophy, or cancer." She praised Thorazine as a "wonder drug," wrote of the tricyclics in the same terms, and announced that lithium, while having "almost no side effects," left those with manic depression "returned to normal" and, if taken religiously, free of relapses.

Kay Redfield Jamison, a Johns Hopkins psychology professor and MacArthur award-winner, published *An Unquiet Mind*, a memoir of her own manic depression, in 1995. She rendered the highs and lows of her condition: flights of vanishing shyness, the sudden "power to captivate," the ceaseless "desire to seduce and be seduced," the intellectual omnivorousness, followed by crashes featuring psychotic hallucinations. In some cases, manic depression comes with psychosis in the very high or low phases. These departures from shared reality tend to be temporary, unlike Caroline's unremitting voices. But they can be no less haunting. Jamison had blood-filled centrifuges spinning and clanking inside her head.

Jamison's narrative was riveting, but the undertow of the book came from her reluctant acceptance of medication, her regretful embrace of the idea that she would be taking it for the rest of her life, her candid descriptions of what lithium had cost her. The book made a mockery of Andreasen's good cheer. There were years with attacks of "severe nausea and vomiting many times a month," but this side effect, which abated with adjustments in how her lithium was administered, was the least of the drawbacks that accompanied the drug. She had always taken pleasure and pride in being an athlete, in playing squash and riding horses. Her skills were now gone forever, lost to the drug's assault on balance and coordination. Her intellect was in jeopardy. The drug impaired concentration and left her unable to comprehend literary novels or book-length nonfiction. She "took up needlepoint as a diversion and made countless cushions and fire screens in a futile attempt to fill the hours I had previously filled with reading." She addressed her medicated transformation in a list of reminders to herself entitled "Rules for the Gracious Acceptance of Lithium in Your Life." It included items like "learn to laugh about spilling coffee, having the palsied signature of an eighty-year-old, and being unable to put on cuff links in less than ten minutes," and "Reread the Book of Job" and "Accommodate to a certain lack of enthusiasm."

Jamison was able to reclaim her intellect. Her memoir itself was testament to that. And there was a scene halfway through *An Unquiet Mind* that stood as a kind of thesis statement, a banishment of all uncertainty about the right path. In the scene, she had recently resigned herself to her own treatment, and, as a clinical psychologist, she was called to a hospital emergency ward, where a bipolar patient of hers was in the midst of a manic and psychotic episode. A year earlier, during a therapy session in her office, he threatened her with a knife, was committed to a psychiatric facility, then released, and now, on the emergency ward, he was strapped down—bound by the ankles, wrists, and across his chest—to a gurney. He was screaming in terror, a "primitive and frightening sound," and a psychiatric resident had just injected him with Haldol to subdue him. His bloodwork showed that he had quit taking his prescribed lithium.

"Gradually the Haldol began to take effect," she wrote. "The screaming stopped, and the frantic straining against his restraints died down. . . . As his psychotherapist for years, I had been privy to his dreams and fears, hopeful and then ruined relationships, grandiose and then shattered plans for the future. I had seen his remarkable resilience, personal courage, and wit; I liked and respected him enormously." But she couldn't persuade him to stay on lithium. She couldn't coax him to accept what she herself accepted; and his disorder, maybe more extreme than hers, "cost him his life." She didn't say how. That wasn't the point. The passage ended with the line "We all move uneasily within our restraints." It was a line that pointed ahead to the book's final pages: "There is now, for me, a rather bittersweet exchange of a comfortable and settled present existence for a troubled but intensely lived past."

Still, the last pages weren't without Andreasen's optimism. There was "a wonderful kind of excitement in modern neuroscience," a "moon-walk sense of exploring and setting out for new frontiers." Jamison declared that "the pace of discovery" was "absolutely staggering." Her disease was surrendering its secrets.

Andrew Solomon's *The Noonday Demon*, winner of the National Book Award in 2001, was just as complex as *An Unquiet Mind*, though in a different way. It was, in part, a memoir of Solomon's excruciating journey with unipolar depression, a journey of alarming descents and arduous climbs back up and then deeper descents, with none of Jamison's ecstatic heights. But where Jamison's slim memoir gained its layers through her sense of loss, Solomon's six hundred pages accumulated insight through the author's expansiveness—through an inquisitive energy that ran counter to the paralyzing force of his condition and that took him as far as Cambodia and Senegal to find out what non-Western cultures do to help the depressed.

In Dakar, Senegal's capital, Solomon underwent an *ndeup*, a rite to purge the malign spirits behind mental illness. His naked body was painted with a sanctified sludge of millet, he lay under multiple blankets while spooning a ram, and he was led in a frenetic dance and finally covered in the blood of the ram and a rooster. "The *ndeup*," he wrote, "provided a way of thinking about the affliction of depression—as a thing external to and separate from the person who suffers," a wicked spell to be expunged.

The allure of *The Noonday Demon* grew from Solomon's intellectual curiosity and geographical wanderings, yet the book was biomedical at its core. "I'd love to see fewer side effects," he wrote about antidepressants, "but I am so grateful to live in this age of solutions." For his treatment, he adhered to psychotherapy alongside medication, but the book skated past the therapy while expounding on the drugs. Solomon warned his readers against the experiment of suspending their use once they'd been effective. It was an experiment he had tried. Much smarter, he said, to think of medication as lifelong. He quoted an eminent biomedical psychiatrist who admonished that relapses of depression were harder to treat than first episodes. "If you have a relative or a patient on digitalis, would you think of suggesting he go off it, see if he has another bout of congestive heart failure, and have his heart get so flabby that it can never get back into shape again? It's not one iota different."

Since his first visit to a psychopharmacologist seven years before writing *The Noonday Demon*, Solomon recalled, "I have been on, in various combinations and at various doses, Zoloft, Paxil, Navane, Effexor, Wellbutrin, Serzone, BuSpar, Zyprexa, Dexedrine, Xanax, Valium, Ambien, and Viagra." He continued to take many of these. The Zyprexa, primarily used as an antipsychotic, he used as a sedative and antianxiety agent, though it made him gain considerable weight and sleep ten hours a night. Yet some nights he couldn't get to sleep; for that, he took Xanax. The antidepressants, or possibly the Zyprexa, played tricks on eros. The Wellbutrin and Dexedrine were "to get my libido running again" and to aid with "the universal problem of much-delayed orgasm." The Viagra was to deal with antidepressant-induced impotence.

One more memoir belonged beside Jamison's and Solomon's

among the era's most influential books about psychiatric disorders. But it also didn't belong with those other books at all. *Darkness Visible*, William Styron's 1991 narrative of plunging directly from being honored at an award ceremony to checking himself into a psychiatric hospital, set a standard for candor and literary grace about depression. And it anticipated one of Jamison's and Solomon's goals: "to remove the burden of stigma from mental illness," in Solomon's words. Styron's book had begun as a short piece in the *New York Times*, and in the memoir he explained that the flood of responses made him feel "that inadvertently I had helped unlock a closet from which many souls were eager to come out."

Because of its advocacy against stigma, *Darkness Visible* was—and still is—talked about as a kind of lyrical endorsement of psychiatry's biomedical revolution. The lifting of stigma has long been claimed by biological psychiatry and by the pharmaceutical industry as a result of their outlook and work. To designate and treat psychiatric conditions as medical diseases, to put them in the same category as diabetes or congestive heart failure, is to alleviate their social taint, their shame. The logic is compelling and by now so common that it is hard to see things any other way. I will return to this; the opposite perspective exists among the afflicted. For the time being, it is enough to say that Styron does not fit easily into the biomedical camp.

Darkness Visible contains a thoughtful confusion. It does refer—in a lone paragraph—to depression as resulting from "an aberrant biochemical process" and "a depletion of the chemicals norepinephrine and serotonin." But more often it rejects such clarity. At the outset, about the science of depression, Styron quotes a clinician he admires: "If you compare our knowledge with Columbus's discovery of America, America is yet unknown; we are still down on that little island in the Bahamas." And in the book's last pages he likens scientific theories of depression to hypotheses about "the origin of black holes." He gives no support for medication. He criticizes psychiatry's "stubborn allegiance to pharmaceuticals." He grapples with the long list of fellow artists who've committed suicide, from Hart Crane to Virginia Woolf to Diane Arbus to Anne Sexton, and he refuses anything more definitive than that depression is "mysterious in its coming" and that, with luck, it is "mysterious in its going."

MY BROTHER WAS not a musical prodigy. His is not a cinematic story of early, fantastic talent destroyed by a draconian psychiatric system. His gift was not obvious to everyone, nor to himself, when he was a child or teenager or nearing and passing twenty. But there were signs of special talent. When he was twelve, his piano teacher called a friend, a bassist who, I would learn later, had recorded with Bill Evans and made a number of albums with Keith Jarrett. He had performed with Miles Davis. He was living in Seattle, and Bob's teacher set Bob up to play with the bassist and his band, to audition to study jazz technique with him.

Our parents drove Bob to the rehearsal space, and Bob, all one hundred and five pounds of him, or maybe, by then, all one hundred and ten pounds of him, stood there at a distance, listening to the band tear through an up-tempo version of a jazz standard. Then the bassist beckoned Bob to the piano, and Bob froze. Despite his own penchant for frenzied blues, he was bowled over by the band's velocity. And he was stunned by their musical discipline. The band members couldn't get him near the instrument. But the point isn't his paralysis. The point is that his teacher made the call in the first place.

Into his teens, he wasn't a natural performer. At gigs, with a quintet of high school classmates, nerves could take over and play tricks in his auditory canals. When he sang lead, a bandmate sometimes had to join him at the mic to guide him back on key. By the time he was nineteen or twenty, though, a future in music didn't seem far-fetched. It turned out he could sing quite well, and a renowned baritone, who had just sung his last role at the Metropolitan Opera, and who had starred on Broadway, heard a tape and took Bob on as a student. There were similar hints, at that age, that he might have a career as a pianist. He was also studying dance. He harbored ambitions of becoming a choreographer or composer.

I can't be the judge of his prospects. All I can say, and perhaps all that needs to be said, is that his ambitions weren't absurd or insane, that he almost surely had enough talent to push through as an artist at some level and in one of the ways he envisioned, and meanwhile that his future was as unsure as that of most young artists, those who aren't fortunate enough to be prodigies or to be plucked from the horde of dreamers in early adulthood. He was somewhere on the cusp, the wide, amorphous cusp. He was simultaneously near and far, with no outcome guaranteed, which is what following artistic ambitions tends to mean. You gamble with your life, betting a few years, or betting your entire twenties, or maybe your early thirties. Or you keep pushing more and more years into the pot.

Yet our father did seem to predict, in my brother's case. He had done no such thing with me. I had set out to be a novelist. I was open about my desire with no more evidence that I could succeed than my brother possessed in support of his goals. An honorable mention in a high school poetry contest? A short story published in an undergraduate magazine? But it was to my brother that our father voiced his skepticism. When Bob laid out what he hoped to do with his life, our father stared back at him with unwavering, judgmental eyes, and asked, "Have you reality-checked that?" My brother heard this in high school, in college. "Have you reality-checked that?" And it is likely that he cared more than I would have. Had our father directed that question at me, I might have written it off to his weakness, his limitations, his fears of anything that risked any form of anarchy. With no shortage of arrogance, I would have dismissed his caution. But to my brother, he was the man who cast a gigantic shadow. It was impossible to ignore his eyes and words.

When he was twenty or twenty-one, I bought my brother a gift. I have said that I was threatened by him and violent toward him. The gist of this had not changed, except that by late in high school he had become taller and stronger than me, which put an end to the violence. Something I've suppressed thoroughly yet recall both dimly and undeniably is that, in part, I felt his hospitalization as a kind of victory. I didn't say as much to myself. I didn't let the feeling become conscious. But the feeling was there.

Still, I had grown to be supportive. One foggy afternoon, we had ridden one of the ferries that crosses between Seattle and the islands of Puget Sound, and, on the passenger deck, he pulled off his sneakers and socks. He said that on previous ferry rides he had choreographed a dance to the vibrations of the boats. Now he demonstrated it for me. Barefoot and wearing a Cowichan knit hat with ear flaps and chin strings dangling, he leapt from side to side, crouched, spun, sprang, floated. He pounded his heels against the deck with its chipped blue paint. There weren't many people around, but there were a few who turned from the rail and more who may have been watching from behind the windows. I was embarrassed, momentarily. Mostly, though, I was drawn in. The vibrations of the engine had a rhythm, a cyclical, subtly shifting tempo, a rise and fall in volume, undulations in pitch. This became music as Bob danced. He was doing what site-specific art is supposed to do. He had heard something others didn't hear, not in this way. It sounded like a Gregorian chant mixed with a kind of orchestra, notes constrained but evocative, mixed with a Native American drumbeat from the throb of the water against the bow. And he had felt something others didn't feel: that this music was in direct communication with his body, telling him how to move.

He had only to listen. So the deck became a sanctified space of art, surrounded by fog and the silhouettes of land, and Bob was the agent of transmutation.

I bought him, at a thrift shop, a white tuxedo jacket. This might seem a strange gift for a ferry dancer, but it made sense to me. The secondhand formal jacket, threadbare but flamboyant, had a vaguely punk feel; I gave it in recognition of his risk-taking. He was on a path I was incapable of following. I was a conformist. In my attempt to become a novelist, I had done nothing unusual. I had kept myself as incognito as possible. I had graduated from college, interned at the most mainstream of literary magazines, mutely read through the slush pile of unsolicited manuscripts, taught English at the Brooklyn school we attended as children, and socialized with a set of soon-to-be corporate lawyers. He had walked away from college, and, despite our father's admonition and his own history of freezing up as a teenager, he had found the freedom to perform anywhere. I gave him the shawl-collared white jacket in admiration.

Something else was happening. He worked for months cutting trails in Olympic National Park. The feeling began there. Back then the park attracted few tourists. He was camping at night, clearing slender paths all day, always ensconced within the park's implausible juxtapositions: ocean next to rain forest next to glacier-covered mountains. The Olympic Peninsula was its own unlikely cosmos at the outermost edge of the continent. It was rock archways and surreal statues carved by the Pacific; it was killer whales; it was the towering forest canopy; it was the summer fields of snow; it was a rock promontory, where he slept and felt he was lying in the middle of a sea; it was a burst of something like St. Elmo's fire, except much stronger, that turned the night sky as bright as day. It was all too spectacular to make sense, and so the feeling of a presence wasn't nonsensical. Or, to say it another way, the world was laden with meaning, meaning just beyond articulation, and so one more element of significance, hovering invisibly close by and communicating pointedly yet incomprehensibly to him, was only another part of the dense, pregnant atmosphere.

It, this unseen and not-quite-verbal presence, trailed him back to Seattle when the job ended. It left and returned. Persisted, off and on, for months. Materialized, immaterial, in the apartment he shared, on the streets he walked. He thought, *You're here. I don't know what you want from me.* He wrote a song, addressing it. The chorus stated that he was not ready. It went on pursuing him, without aggression.

He visited the physics department at the University of Washington. He stepped into the department office and said to the office assistant that he had an issue he'd like to discuss. He outlined his thoughts, and she introduced him to three bearded, thirty-something graduate students. The issue, as Bob explained to them, was a conflict between the interconnectedness of all aspects of the universe, a spiritual knowledge he felt viscerally from his time on the Olympic Peninsula, and the fundamental mistake he perceived in particle physics, which aimed to understand the principles of existence by propelling subatomic pieces of the universe toward each other and smashing them together in order to split them apart. He elaborated on this error. The graduate students spoke to him about the concept of superluminal connections, gave him a reading list, and invited him to join their study group.

A belief that he was getting beyond whatever it was that had made him doubt and denigrate himself as a musician, whatever it was that he traced back to his failure to venture so much as a note with the jazz bassist and his band; a sense that he was being followed, that something was being asked of him, that the request might or might not have to do with music or dance, though it was clear that something was in store if he was willing to turn toward it; a confidence that came from the interest of the three doctoral students in his ideas—all of this added up to a feeling that he was on the verge of breaking through in some unknown way.

And then old judgments returned. They rose from their roots, crept upward, and ensnared him. He had the tee shirt printed with the message across the chest. A dark tee shirt. White letters. No distractions. Easy to read. "I feel like I don't have anything to offer." His shaggy hair was damp, his sneakers wet from long walks in the Seattle rain. He wore the shirt along the main avenue of shops in his neighborhood. He wore it to dance class.

People said, "What do you mean?"

People said, "Why would you say that?"

People said, "That's stupid."

Our parents found him a therapist, who sent him to a psychiatrist, who prescribed an antidepressant, a tricyclic. It did nothing and was abandoned. A roommate moved out and was replaced by an alcoholic. A girlfriend left to study in Europe. The presence of the alcoholic in the apartment, the disappearance of the girlfriend from his life—these amplified the truth of the message he wore.

He came across a copy of Salinger's *Franny and Zooey*. The book took hold of him and pulled him ahead at a gallop. Franny dropped out of college. Full of self-loathing, she despised her talent as an actress. She hurtled into a breakdown. In the last pages, she was in bed, listening to her brother over the phone: "Don't you know that goddam secret yet?" he asked her. "And don't you know—*listen* to me now—*don't you know who that Fat Lady really is?* . . . Ah, buddy. Ah, buddy. It's Christ Himself. Christ Himself, buddy." He understood her brother's meaning as immediately as she did. It was cryptic to neither of them. Her brother hung up the phone. Then: "Franny took in her breath slightly but continued to hold the phone to her ear. A dial tone, of course, followed the formal break in connection. She appeared to find it extraordinarily beautiful to listen to, rather as if it were the best possible substitute for the primordial silence itself."

It was nine miles from his apartment to the house where our parents lived and where we'd grown up. Sharing Franny's elation, he set out, walking, after closing the book. He trekked along Lake Union and across the Fremont Bridge and past the locks that let the salmon swim from the sound into the freshwater where they spawned. He climbed the hill to the isolated house with the raccoons on the roof of his room. He told our parents that they needn't worry about him any longer. He was better, absolutely better. He explicated the end of Salinger's linked stories.

Our father had a way of tipping his head forward, minimally, even imperceptibly, and seeming to gaze out of the upper part of his eyes. During television interviews, he combined this gaze with his monotone voice in order to impress on the public what was best for their health. Bob, informing them about his new insight and restored confidence, received something akin to this stare, a look he felt as withering. Our mother's words were in sync with our father's eyes.

He walked back along the locks, over the Fremont Bridge, past Lake Union, and to the apartment he shared with the alcoholic. There, he came to a firm resolution. Lately, he had been chatting with and sometimes giving money to a middle-aged homeless woman who lived with her brother in a tent not far from his apartment. He decided that if, the next time he went outside, he saw the homeless woman, he was going to follow through on the plan that had taken shape in his mind. The woman would be his sign.

He walked out the front door of his building. There she was.

She said she was looking for him, that her tent had burned down. He asked if he could buy her food, and she requested Chinese. They ate together in a Chinese restaurant, where she told him that her brother had been burned in the fire. They visited him in the hospital. Bob insisted on calling their estranged mother, who lived on the other side of the country, and trying to explain that her children needed her badly and that she should send them money. This backfired; it led to a scene. Still, the sign was a sign. Back at his apartment, he packed a small shoulder bag. He phoned Grandma Clara in Brooklyn to tell her he was on the way. He put on the white tuxedo jacket and headed for the airport.

His plan, according to what our parents would soon tell me, was to stop over in Brooklyn, cure our grandfather of his fast-progressing dementia, and then board another plane, this one to France, where he would travel to Poitiers, a town southwest of Paris, where he would draw some sort of sustenance from the spirit of Joan of Arc.

Both he and Joan of Arc had ties to Poitiers. He and I had been born there while our father was stationed just outside the town during an obligatory stint with the U.S. military. For Joan of Arc, Poitiers, the center of French theology in the fifteenth century, was where she was interrogated by clerics seeking to determine if she had heard God's voice. She maintained that she had. She was the illiterate teenaged daughter of a farmer and village official, yet she claimed, first to military officers and now to her clerical inquisitors, that God was directing her to liberate France from English rule, starting with the English siege of the city of Orléans. She demanded to be put in charge of French troops.

"You assert that a Voice told you, God willed you to deliver the people of France," one of her inquisitors said. "But God wills that you should not be believed unless there appear some sign to prove that you should be believed. And we shall not advise the king to trust in you, and to risk an army, on your simple statement."

To this, Joan replied, "I am not come to Poitiers to show signs. But send me to Orléans, where I shall show you the signs for which I am sent."

The clergy and king sent her. She and her soldiers freed Orléans. In the Hundred Years War, this changed the fate of France. Bob saw meaning in the coincidence of his birthplace being the town where Joan was bombarded by doubt, stood up to the skeptics, won them over, and rode off to prove she was right about the voice she had heard. He wanted to be in that place. No matter what path he ultimately took, he would be filled with strength by being there.

Our parents, told by Clara that he was on his way to visit her, and alarmed by his lurching moods and impulsivity—they didn't yet know about his larger plans; that came over the next day—tried to intervene without igniting him. Instead of chasing him to the airport, they managed to convince the airport security team to send someone to intercept him at the gate. This was another era; security had time and energy for such things. A Sea-Tac Airport agent took him aside. Bob was too old to be considered a runaway, so there wasn't much the agent could do besides have a conversation. He asked if Bob knew that his parents were worried. He asked why Bob was wearing a white tuxedo jacket.

At the other end of the flight, at JFK, our parents had a family friend meet him and cajole him gently to her son's apartment. Our parents were waiting for him there. This is one part of what happened that makes no sense to me. They hadn't chased him to the Seattle airport, yet they reached New York before him. Perhaps he hadn't been on a nonstop flight. I pause to dwell on this sometimes, only because the fact that they awaited him must have contributed to his fear—already forming—that he could not escape them.

Our parents gazed at their son. It was November; he may have had on his wool hat with the ear flaps, woven in a Cowichan pattern, clashing with the secondhand formal wear. Either that or his unkempt hair was exposed. Their hearts were breaking. They proceeded with delicacy, like diplomats, careful to speak evenly and incrementally, to say nothing that would cause the discussion to blow up. He didn't trust them but started to think that— because he wouldn't see them again for a good while, a necessary precaution if he was going to accomplish whatever he was meant to bring about—he could afford to humor them. He was also in tears. He agreed to fly home with them for just the weekend. And then, as their incremental strategy progressed, he agreed, at some point during the next several hours, to spend the weekend in a hospital, just the weekend, and just to appease them, and only if they pledged to buy him a replacement plane ticket to New York and on to France as soon as this two-day delay was over. On the plane to Seattle, he sobbed over a poem in a magazine. At the hospital, he voluntarily committed himself. That he did this, that he consented to his own psychiatric incarceration, would soon compound his feeling of our parents' inescapable control.

"HE DECIDED TO go to France based on some prenatal experience he had there." This is the only explanation the psychiatric staff recorded for his leaving Seattle. And it is possible, though I think not likely, that neither Joan of Arc nor our grandfather's dementia played a role. Much later, he didn't recall either as a motivation, though he did remember Joan being a compelling figure in his life. He recalled going to Brooklyn for the purpose of saying goodbye to our grandparents, whom he thought he might never see again, because he expected to live in France indefinitely, after rectifying the prenatal experience alluded to in the records. This experience, endured in the womb, was the failure to ride a cable car up to a peak in the High Alps. He'd heard the story from our parents. He interpreted it as seminal. Our father and I rode up; our mother and Bob remained down in the village, because the altitude was deemed ill-advised for pregnant mothers. A first step in being sure that his life was on the right course, and in bolstering the confidence and defiance he would need to stay that course, would be to reach that peak.

Communing in some way with Joan of Arc in Poitiers was a motive told to me by our parents at the time of Bob's hospitalization and by my brother shortly afterward. About the motive of rescuing our grandfather from Alzheimer's, I'm less sure, because I don't think I ever heard it from his lips. Only our parents said that it was one of his reasons. Could they have invented this in their distress? Taken something he'd said and turned it into this? It is possible. But I don't think they did invent it. The belief that he could cure our grandfather seems unlikely to have been a product of their imaginations.

The hospital records are general. "Claims special powers," they read. They refer to his "flight of ideas," his "grandiosity," his "ability to read people's minds and predict the future," his "receiving messages." They get no more specific than that.

What is certain is that he woke before dawn after spending his first night on the ward and saw his situation in a new light. He stepped out of his room and recognized, to an extent he hadn't been capable of when he arrived, that the ward was locked, impassably locked, that he was locked in. He felt that he'd been tricked, by our parents, by the ward staff, tricked into signing his rights away, signing *himself* away.

What rights did he have left? Did he have any at all? The most basic one? He had no idea. He decided to find out immediately. At five in the morning, he walked over to a nurse and demanded to be let go. She called the psychiatric resident. Bob picked up a phone and called Air France and began trying to book a ticket. The resident called for more staff. Someone on the ward called our parents, and our mother appeared.

"All that talk," she told him, when he reiterated some of his artistic plans, "about being a musician, a singer, a dancer—that is a sign of your illness. If you leave here now, we'll have to scrape you off the sidewalk."

He stayed. He acquiesced to medication. He wasn't sure they needed his consent, and after our mother's lecture, he felt that his will wasn't relevant to anything. Three psychiatric drugs spread through his brain. Judging by the hospital records, one of the drugs, Mellaril, an antipsychotic, was given mostly to sedate him. This drug he was fed on a regular basis. It accompanied a regularly administered and high dose of Haldol, though by 1983 Haldol's frequently destructive and often irreparable side effects were well established. The staff, in the hope of mitigating this danger, but probably well aware that it could not be sufficiently reduced, added a non-psychotropic anti-Parkinson's medication, amantadine. The third psychiatric drug was lithium, to stabilize his mood. It was given at a dose aggressive enough to bring his blood readings into the upper therapeutic range.

"No suicidal or homicidal ideation," his records state. Yet to keep him groggy on the ward, to eradicate his belief that he was receiving messages, to make sure that his moods stayed within common bounds, and, inevitably and above all, to abide by protocol and eliminate the presumed if nonevident risk of suicide, he was infused with these medicines.

Quickly he learned compliance. He did and said what the staff and our parents wished him to do and say. This entailed not only swallowing his medication obediently but also taking part during sessions of group therapy. When other patients spoke darkly, he countered with light. When they spoke with futility, he said, "When you're ready to get better, you will get better." He told the staff that he planned to return to college and listed possibilities for work until he could re-enroll.

The antipsychotics were removed from his regimen. This is typical in the treatment of bipolar disorder, as the temporary psychosis of an acute manic phase subsides. There was nothing unusual about anything the hospital staff did with my brother, not by the standards of the 1980s, not by the standards of today. He was discharged after two and a half weeks, with the understanding that he would be on lithium for the rest of his life; with, as the records note, a tremor in his hands; and with, as he told me, the feeling that he had a blanket over his brain.

SEVEN

T he farm was just past a one-stoplight town in southwestern North Carolina. Coming from the suburban landscape of Zionsville and the campus environs of Bloomington and the grounds of the Houston hospital with its IKEA trees, this, in the Appalachian foothills, was far from anything Caroline knew. The staff put her through a series of daylong trials with a number of work crews, to find out where she might fit. There were kitchen details and greenhouse crews and teams that tended fields of crops. On each of her first days, she tried to take part in what a new team was doing. She wore her black uniform—her Bikini Kill or You Can't Rape a .38 hoodie, her *Cabaret* wig. Her voices screamed, scared by so much that was unfamiliar. They warned that motherfuckers lurked here who wanted to kill her. The screaming was so persistent she couldn't hear the instructions of the crew leaders.

Nights, she saw the stars. This felt like an extravagance after her months locked up at the hospital, where the cage was the closest she got to outside. Mornings, she swallowed multiple antipsychotics and other pills under a nurse's vigilant watch, struggled to concentrate amid the voices that did not defer to the drugs, and attempted to imitate yet another team. The staff decided to put her on the animal crew. The joke of the cage was complete. Her punk ass was assigned to tend the sheep.

She pulled herself into the open bed of the fat-wheeled orange Kubota that took the crew from the main building to a red barn. Two mutts hopped into the Kubota bed, too. There were chickens to feed, eggs to collect, animals to water, shit mounds to shovel, stalls to clean. She climbed the ladder to a loft, twisted her body, and flung hay bales down. She discovered that she was strong enough to hoist and haul fifty-pound bags of grain.

There was an irascible horse that bit people's hands. She was a miniature, the top of her tawny head reaching no higher than Caroline's shoulders. Because of her biting and because she was probably going to get herself banished to another farm, Caroline bonded with her, though even after their friendship was established, the horse kept on clamping her teeth down on Caroline just as hard as on the others.

The sounds of the crop teams calling to one another rose from down the hill, rose from fields of potatoes and radishes and kale. In shoveling shit, in stroking and scratching the mutts as they rode in the back of the Kubota, in stroking and offering solace to the biting horse, who could be soothed only fleetingly, in talking to the chickens and lifting feed bags to her shoulders, weeks went by, months went by. Her voices shed some of their fear, hostility, volume, density. The sounds of the crop teams and of the llamas in the lower barn could more easily slip between them. She could hear instructions almost clearly.

One morning, the crew reached the barn to find that a ewe had given birth in the night. The lamb's leg was broken, stomped by its mother in the chaotic moments after delivery. Below the break, the leg swung like something that didn't belong to the body, something stitched on by a few internal threads, an anatomical afterthought. The newborn looked crippled, doomed, a victim of nature's indifference, its arbitrary cruelty. With a coat the shade of chocolate milk and yellow eyes in a prodding, poking head, it seemed oblivious to its fate. The staffer in charge asked Caroline to help set the leg, to wrap from below to above the break in pink cotton tape, snugly but not tightly, taking care to allow for circulation, then to cut a section of white plastic pipe and splint the leg.

"Someone's going to have to feed her," the crew chief said.

There was formula in a supply shed, the domain of a hefty, rugged cat. Caroline mixed a feeding's worth in a bottle and, back in the barn, found a shit-free spot on a bed of hay. She knelt down and steadied the lamb against her body. Once it was sucking formula, its tail wagged at a dazzling speed. Within seconds, the lamb seemed drugged, electrified. It couldn't keep the nipple in its mouth; milk spurted into its eyes, splashed its spindly neck. Caroline had to learn to calm her with her touch and body and voice, to keep her focused on what she needed. It was strange, she thought as she gained this expertise over the coming days, strange that here were residents who couldn't calm themselves, residents who heard voices like she did, residents learning to calm and control the animals, to guide them into the right pens, coax them onto a table to be shorn, keep milk flowing into a mouth.

The staff decided it was time for her to quit smoking. To assist in this, they added a medication, Wellbutrin, an outlier among the antidepressants, a pill thought to work not by blocking the reuptake of serotonin but by boosting the brain's dopamine as well as norepinephrine. For some, it seems to bolster mood; for a small minority, it supercharges anxiety; for others, it does nothing. For smokers, it can dull cravings for nicotine. In Caroline, it reversed what the miniature horse and the lamb with the splinted leg had done for her. Her voices resurged. They populated the barn, awaited her in the main building, stood at her shoulder in the dining hall. The one she'd known the longest, who'd once warned about the threat to her family, told her who at the farm intended to kill her and instructed her, screamed at her, to do the killing first. Partly following and partly flailing against his commands, she tackled a resident. At dinner, she threw chairs. "Dude," another resident asked her, "is this because of the shit they put you on to quit smoking?"

By this time, she had graduated to a level of independence with her medication. The staff still watched over her, but to a lesser degree. She no longer swallowed her pills at the med-room window, mornings and evenings. Instead, once each week, at the window and under watch, she allotted her many drugs into a many-slotted plastic pill case, with a staffer checking that each pill was in its proper compartment. She was permitted to swallow her medication in her room.

The staff eliminated the Wellbutrin. And Caroline began to contemplate something more. She had no idea how to stop the other drugs, except to cut them cold. And she had no idea what to expect if she did. All she knew was that benzo withdrawal had brought on a grand mal seizure and that Wellbutrin had almost made her a murderer. She wondered what the rest of her pills were doing to her system. Reading online wasn't possible; at the farm, all accessible computers were in strategically public places. She couldn't risk being seen or having her searches checked.

She had a new companion. The barn, under its double-slanted roof, was vast. She greeted the sheep and her healed lamb, greeted the rabbits and the splenetic horse, on her slow, eager way to Apona's stall. "What's up, Apona? What's up, girl?" Caroline said. Apona was a runtish, aging donkey with a middling gray coat. She had a bushy mane that defied all attempts at grooming. "Let's get you going. Let's get your manicure going."

The ritual was this: She went to a shelf of rough wood beside a latticed window. She took a jar of rich black keratin liquid and a brush into Apona's stall. The donkey lowered her broad snout and straightened with a few blades of hay in her mouth, chewing absently. "You ready, girl? You ready for your manicure?"

Apona's hooves were chapped and peeling. Caroline knelt on one knee. She needed to be in a particular position in order to treat the hooves with the precision and tenderness that was her goal. This was because of her shaking. It was unpredictable. Sometimes it was just a quivering; often, though, she had a hard time lighting her own cigarettes. At some meals, she couldn't manage a fork. She hoped no one would mock her for the mess she made, that the trembling and jerking of other residents obscured her own.

Below Apona's pendulous belly and beside one of her legs, Caroline arranged herself in a half-kneel, half-crouch, clenching her body and all but compressing herself into a ball, her chin on the upraised knee. Otherwise, she felt that her head would tug itself out of her control, to start tipping and flinching and twisting. She braced the forearm of her brush hand against her shin. Long ago, she'd had a fantasy: she with friends, painting each other's nails, sharing delicate, undistractable attention with pretty bottles and dainty brushes. But even if she'd had the friends, her quivering would have caused disaster.

She dipped the brush into the jar and held one of the hooves. It was a chalky, fibrous gray, but as she painted horizontally, rounding the curves, the surface darkened, gleamed. She returned the bristles to the spot where she'd begun and repeated the painting. Apona dropped her head occasionally, plucked up another few blades, and chewed without hunger, as if to communicate her satisfaction. The hoof took on a black luster.

"You're looking good, girl," Caroline said, when all four hooves were ebony and resplendent and she couldn't linger in Apona's stall any longer without neglecting the rest of the animals. Apona gazed at her, continuing with her chewing. "See you later, girl."

Staff were beginning to say the same to Caroline, that she was looking healthier. Six months into her stay, she was losing weight.

Her dark hair was thickening. Some mornings, she left her wig in her room.

"How's your med-packing going?"

She had progressed to another level. She didn't have to pack her pill case under anyone's watch; rather, she was given the bottles and allowed to distribute the pills into their slots on her own, in private, in her room—trusted.

"Going great."

"No problems with the packing?"

"No, no problems."

They told her country living seemed to agree with her. They spoke with pride about the effects of farmwork and a homegrown diet. She agreed, sincerely; it couldn't have been more true. But another factor was that she'd stopped taking her pills, all of them, the antipsychotics and the rest, all at once. She flushed them down the toilet, morning by morning, evening by evening, careful that if anyone checked her case it would be impeccably up to date, exactly the right number of pills remaining.

"How is everything, Caroline?" the staff psychiatrist asked.

She told him all was well.

He proceeded down his checklist. Thoughts of self-harm?

No.

Thoughts of suicide?

No.

Thoughts of harming others?

No.

Hearing anything strange?

No.

Her voices were, in fact, more talkative but less combative, less prone to wordless cries, less likely to urge her to strike first, murderously, or to coach her on methods for ending her own shameful life. There was also a paradox: Off the medications, her voices didn't collide and overlap and become unintelligible, or speak as if through heavy cloth, and their clarity was easier to cope with, even when they did ridicule her or demand violence. The rest of the world, beyond these unseen, audible presences, became more fully perceived, more vibrant: the immense, asymmetrical trees; the fairy-tale bridge arching over a stream; the ragged banks of a pond; Apona waiting in her stall; a reedy, inward girl, a new resident with unruly, quaking hands, whom Caroline mentored in the barn, teaching her to manicure Apona's hooves.

CAROLINE GRADUATED, AFTER a year, to a group home. The staff sent her down the road to Tryon. Asheville, the other possible placement, was judged too dangerous. She'd made such progress, taking ownership of the animals and taking fellow residents under her wing and taking charge of her medication; still, Asheville was a city, and she had a history of substance abuse.

Tryon had a population of sixteen hundred, a statue of a wooden horse to commemorate the town's bygone toymaking factory, and, on its quaint commercial street, not a job to apply for. Caroline turned the fan in her room to its highest, loudest setting, and wept. It was one of her two main experiences in Tryon, expelling her misery into the muffling wind.

The other was learning about Nina Simone. She heard Simone mentioned in a coffee shop and saw the singer's picture on the coffee shop wall. She read about Simone online and walked away from downtown, the streets growing more and more hushed, to find her childhood home. A mile outside Tryon, it sat, abandoned, on a hill of mangy grass and dust: splintering wood slats bare of paint; the remnants of a railing that had once bounded the steps leading up to the front porch; a round metal sign nailed beside a decrepit window, using Simone's given name and reading only "Eunice Waymon" and "Birthplace"; a back entrance whose stoop was altogether gone, replaced by tall weeds.

During Caroline's time in Tryon, Simone's music became a much better option than the fan's blast. There was the stripped-down, ferocious version of "I Put a Spell on You," with its wild blurts of rageful scat singing. There was the swell of a gospel choir behind Simone on "Young, Gifted and Black," turning her lines— "When you feel really low / yeah, there's a great truth you should know"—into a towering wave. No matter that she wasn't gifted or black, Caroline felt the pounding crush and thrall of Simone's singing.

At last, she moved to Asheville, to a group home on a dead-end street of rough pavement and unsold lots. Well-practiced from the farm, she replied to the staff's questions about packing her case and to a visiting psychiatrist's inquiries about odd or violent voices or thoughts. With discipline, she dispatched her pills into the sewer system. A toddler, a boy, joined the other voices. She caught a bus by a Family Dollar and filled out job applications at clothing shops, diners, and a place that baked holistic French desserts for pets. She was called for no interviews but found volunteer work with a crisis ministry that gave clothes away. She decided to approach everyone as if they were browsing at a high-end department store. "Oh, I just got something in," she told the shoppers, who came with their vouchers, "and it is going to look great on you."

It was October 2008, and one of the house social workers pressed the residents to come with him to hear Obama speak. The social worker ushered them into the group home van. Climbing into that vehicle always made Caroline shrivel: being loaded in; being semi-human; being cargo. They drove to a high school that looked more like a college, with a brick archway and cupola, a grand tower and a lavish football stadium. The stadium was jammed, mostly with people around her age. They waved signs reading "Change We Need" and awaited the Democratic nominee, who sprang onto the stage, in a crisp white shirt and no suit jacket, radiating confidence that he would make history. Three months later, watching his inauguration, tears streaked her face and another resident broke out in hives as soon as he spoke his first words, "My fellow citizens." Because this day was so unprecedented, because his odds had been so long—as long as theirs at just about anything, as they sat in front of the TV in that group home on that dead-end street—she felt the reach of those three words run through her like a current.

Next, the weekend social worker took them to a roller derby bout. This meant another ride in the white twelve-seater that she loathed. There was no sign painted on the side, but Caroline felt that the bulk and color of the vehicle advertised that mental cases rode within. She dreaded the moment of being disgorged and walking across the parking lot. She was shaking less nowadays and had lost fifty pounds, but she suspected that, to any outsider, she was indistinguishable from her housemates, who fought errant movements and whose bellies and plumber's butts peeped out around ill-fitting clothes. There was something, she sensed, unmistakably downcast and mortified about all of them. They broadcast their past stays on psych wards, their current status under supervision. And if she had any doubt about being recognizably one of the misfits, one of her voices specialized in reminding her. This was Miss Kathy, who'd been at her side since middle school, ordering her not to raise her hand. In the van, she told Caroline she was unclean, that a stench rose from her skin, that she shouldn't deceive herself into thinking she belonged, even for a few hours, in the outside world.

About roller derby, Caroline was vaguely intrigued but expected to be disgusted. She anticipated short-shorts and push-up bras. She pictured slinky, buxom women skating around a track in a game that was somehow scripted, bumping bodies and taunting each other into pseudo-brawls, providing a crowd of men with titillating violence and titillating flesh. But taking a seat on the lower level of the small arena, she saw that it might not be like that. The crowd was equally women and men. Families were scattered through the sections. Below, on an oval track outlined in tape on the gray concrete floor, skaters warmed up. They weren't wearing short-shorts or Victoria's Secret. The home team, the Blue Ridge Rollergirls, wore tanks in black and blue, black workout shorts, heavy pads, and old-school, ungainly helmets. And slinky wasn't their dominant body type. The team was burly, sculpted, slim, thick-thighed, tall, short, knifelike, blubbery-but-scary. The skaters were white and black. They bunched themselves in packs and practiced unified, militaristic movements. Then the visitors took their turn warming up as the home team skated off to screams.

The bout looked like rugby, minus the ball. Skaters bent double in wide stances, maintaining their poised, weaponized postures until a striped referee blew his whistle twice. Then the teams straightened, tangled, drove at each other with lowered shoulders and sent each other skidding to their knees or spinning onto their asses. They realigned themselves into walls of enforcers working ahead of fleet, nimble teammates, who tried to slip along the sidelines or slither through the horde.

She started to catch on. The crafty, fast ones, who scored points by flying past opponents, were called jammers. Outright endangerment put you in the penalty box and gave the other team a power jam. But as long as you kept your elbows down it seemed that hard, hockey-like body checks were on the right side of the rules. There wasn't much that was pretty about the sport—though the jammers did glide and dance—and the unprettiness was part of what appealed to her. So did the strategies, the formations, the aggressive collaborations of the women. And so did some of the names yelled out by the fans and spoken by the M.C. who called the bout. Sugar Magmaulya was one of the skaters. There was humor but not in the way the name was yelled or spoken.

The next day, Caroline bought skates, Riedells, black, with

the scent of new leather and the sound of new bearings. Because of the catcalls and come-ons that seemed a routine part of being female and waiting for and riding the bus in Asheville, she had switched to commuting to her volunteer job on a fixed-gear bicycle, built for her by a pair of staffers who were amateur racers. There were hills to climb between home and work; her wind had improved and her legs were sturdy. She took the skates to an outdoor rink in a city park, practiced for a week, and decided she was ready. She'd learned that the team drilled at the same rink. One afternoon, she unlatched the door in the boards, skated straight up to a pack, and said, "I want to do this." The desire leapt out of her mouth that way: direct; devoid of superfluous words. No one laughed. No one said no. They said, "Oh, girl, those wheels are garbage" and "You're going to have to learn how to fall."

This she did, for months. She practiced solo; she was pummeled by gauntlets. She learned to slide on her padded knees and glide back up. She showed a certain talent: her natural speed as a skater wasn't bad, but it was the rising, whether out of a skid or out from beneath a bigger woman, whether after a sideswipe that slammed her down or a scoop that lifted her off the rink before she landed on all fours—it was the way she regained her stride, as if she'd never actually lost it, that caught everyone's attention. Sugar Magmaulya became her derby wife. As a newbie, Caroline was automatically Fresh Meat. In the sport's slant on prison language, this didn't signal that she and Mag were a lesbian couple, only that if the time came when Caroline was injured in a bout, broke her arm or needed stitches, Mag would take her to the emergency room instead of going to the after-party.

Back in the smoking cage, Caroline had found camaraderie, but it had grown from diagnoses and psychiatric incarceration. At the farm, she'd woken and willed herself past the denigrations of her morning voices to get herself into the Kubota and reach the companionship of the animals. This, with the team, was new. It was human beings and wasn't rooted in fragility. It flourished in bruises and a slang of bravado. It was a veterinary technician, a nurse, a lawyer, a woman who worked with survivors of domestic abuse, another who worked in a hardware store. It was moms. It was women who were both: freaks who had these everyday roles. Sitting with her two pit bulls on her front porch, Mag talked about her dream of becoming a pastry chef.

The time came for Caroline to have her own name. On the patio of a bar with a group of skaters, she met someone from another team, a black woman who went by Scarriett Tubman and wore the number 40+1, a reminder of the forty acres and a mule promised to freed slaves but never given. The skater seemed to be saying, I'm here as a different kind of Harriet, ready to do damage and take what's mine this time around. With her, Caroline laid out her own situation: nutcase; group home; jobless; the one thing she took pride in, when she wasn't on skates, being that she was Jewish. Scarriett Tubman didn't miss a beat. "You need to be Mazel Tov Cocktail," she said.

That was who she then was, the jammer known as Mazel Tov Cocktail, never Caroline to her teammates, never Caroline to the derby fans crowding the three-thousand-seat Civic Center as she tipped her toe stop against the jam line and awaited the referee's double whistle, releasing the skaters and starting the bout. She pried at the opposition pack and braced for the collision with bodies much broader, thicker, taller than her own. She leaned into the jolt. She lost control of her legs as they scissored with another skater's. She gyrated, was slung to the track, kept her limbs from splaying, straightened as she slid. She veered to the inside, stutter-stepped along the infield boundary, veered again and wedged her hips and shoulders through the last of the opposing team, with the referee pointing, indicating that she was about to score. She finessed a special blockade—the truck and trailer. She grasped the hip bones of a teammate, who dropped her center of gravity and made herself as solid as a post even as they hurtled forward. She whipped herself off those hip bones, accelerating, attempting to crash through. Upended, her skates went out from under her, and she fell helplessly, painfully backward. There was no way to right herself smoothly. But she clambered fast.

Her photo in the program showed her in profile, hair cut to her jawline, right arm raised behind her head, flaming bottle of Manischewitz wine in her throwing hand. This was the page her fans held open as they began requesting her autograph, as the requests multiplied, women, girls, boys, occasionally men leaning over the rail, which was actually a metal barricade in the make-shift way the arena was converted for use in derby bouts. In the stands, tweens and teens lifted homemade placards with her derby name and exclamations of love and devotion.

The billboards appeared: the team, with her featured in the foreground. She was turned half-sideways, knees bent. She was semi-squatting and prepared to spring. Her helmet bore the jam-mer's blue star. That was her above the Bojangles. That was her above the highway.

BETWEEN WARM-UPS AND the double whistle, Caroline skated over to a client sitting low in the stands. "We're going to whup these bitches' asses tonight," she said.

"Fuck yeah," the client said. "Hit that blond one for me."

Caroline had a paying job. The state had begun funding clin-ics and agencies to hire peer-support specialists, people who'd lived through their own troubles and might help to cut state costs by bonding with high-risk clients and lowering rates of relapse, attempted suicide, hospitalization. Peer specialists were supposed to help monitor compliance with psychotropics and steer clients away from substance abuse. Caroline's training had been run by a middle-aged gay man, raised a Southern Baptist, who'd been devoted to his church—and been broken by the congregation's reaction when he came out. His features were faintly Buddha-like. He talked about how to listen, how to keep in mind the forms of oppression that people faced. To Caroline, it seemed a miracle that the job of peer specialist existed, let alone that this was the knowledge it required.

Her boss, a social worker who ran a large clinic, put a premium on different skills. Maximizing payments from Medicaid was number one. Coming down hard on compliance, making sure you didn't get played, and reporting on anyone who said anything that so much as hinted that they might stray from their programs—these were right behind. To him, Caroline was like an auxiliary cop.

"She's a frequent flier," he liked to say, before sending Caroline off to meet a new client. This was his way of warning that the client had a record of repeated hospitalizations or stays in rehab. Or he coached, "Don't let these people manipulate you." Or, "This one's a classic welfare queen."

Caroline's uncle, who lived in Tennessee, had given her an old Buick, and she drove it up dirt roads into the foothills to find her new assignments. Her boss said that Evelyn craved attention and had a talent for overdosing. Angel statues stood outside Evelyn's trailer. The interior was dim. It was decorated with porcelain figurines representing a girl's growing up, one cute statuette for each year from toddlerhood to teens. Heavyset and in her thirties, Evelyn favored tee shirts with floral embroidery, and Caroline sensed, behind the flowers, the figurines, the angels, wells of shame.

They got to know each other slowly. They talked on a couch, in the light Evelyn kept low. Caroline was conscious of the advantages she'd had over most of her clients; still, she had stories to tell when a story was needed. And she knew when to say nothing except "That fucking sucks," instead of supplying some tidy insight or advice. She knew when to let silence sprawl. One day, she asked

Evelyn if there was anywhere she'd like to go, anything she'd like to do. Her client said she'd like to walk around Wal-Mart. This became a ritual. On one of these outings, as they walked past the Wal-Mart curtain rods, Evelyn said, "My dad used to beat me with one of those."

Jackie lived on a street of boarded-up windows, with a dilapidated four-wheeler in her front yard. When Caroline asked if she'd like to go anywhere, her desire took them no farther than a nearby park. They wound up on the swings, next to each other and pumping high, pointing their toes at the treetops. Jackie asked for tales of roller derby. As their legs dangled and their swings lost height, Jackie said she had two dreams: to get her kids back from child services and to get a car, so they could drive across the state and see the ocean.

A Vietnam vet everyone called Big Al was the clinic's other peer specialist. Part Cherokee, he had white hair and a white beard, small eyes, and discolored indentations on his arms where shrapnel had hit. He wore a leather vest over short-sleeved button-down shirts. He drove a Harley to visit his clients. He scared Caroline until one day when they wept together behind a shut office door, sharing their moments with clients, their own memories, the residual smell, for him, four decades later, of fast-rotting bodies, and sharing their dismay that what their job really entailed, despite the relationships they tried to forge, was getting clients to pee in a cup to be sure they weren't using and filling out a checklist covering suicidality and obedience about medications. Such were the people they strove to know: through urine and a questionnaire.

Her clients sometimes asked her to join their sessions with the clinic psychiatrist. She saw psychiatry at work when it wasn't working on her. In Zionsville, at the Houston hospital, at the farm, she'd always had the benefit, at least, of the attention that private payments and private insurance bought, if this could be called a benefit. Now she saw what came with Medicaid: a practitioner who hardly looked up, who droned as he read the requisite questions, who recorded answers before the client paused. Then he determined whether medications should be maintained or switched for similar pills, based, she couldn't help thinking, on the pitches the pharma reps gave. These women, many no older than Caroline, arrived at the clinic wheeling suitcases of psychotropic samples behind them. They laid out lunches in the conference room. In spiked heels and tight skirts and tops, they performed cheerfully, paragons of well-adjusted existence, reciting the data their companies had drilled into them.

Caroline wasn't sure whose side she was working on, Evelyn's and Jackie's or the clinic's and pharma companies' and the state's. Then layoffs at the clinic led her to a new job, still in peer support but deeper into the territory on the other side of the divide, at a psych hospital. By now, she lived in her own apartment, with Spanish floor tiles, a bathroom scarcely bigger than an outhouse, and sporadic heat in the winter, in a gabled building that had once been a TB sanitorium. Between the apartment and the Buick and the fact that she no longer had to bother with the charade of filling a med case and flushing her pills twice daily down the toilet, she felt liberated—except for the undertow at work, her uneasiness about who she was on those locked floors. She carried the keys.

She led group sessions, trying to impart to the residents the slow rhythms of simply listening—with no one correcting, with no advising unless advice was asked for. She sat one-on-one with residents on the soft chairs in the common rooms, hoping to replicate what had happened in Wal-Mart and on the swings. But while she'd had an amorphous authority in her previous job, here, in a facility the residents couldn't leave without an escort, authority was everywhere in the atmosphere, and she was part of that pressure, that atmospheric weight. She was the air residents were forced to breathe.

She fought this feeling off. She reminded herself that she was peer support, that the lesson of her presence was that the residents could free themselves. Then Margaret was admitted, a young woman Caroline had lived with at the group home. There, they'd been friendly; here, they could muster only echoes. "Dude," they greeted each other. Weeks after arriving, Margaret drank cans of paint in the art studio, and Caroline drove to the emergency room as soon as she heard. She found her old housemate lying on a gurney in a hospital corridor, out of danger but without an available room, a white-sheeted vagrant, with nurses, custodians, doctors, visitors walking past.

"Dude, I didn't even know it was like this."

"Dude."

"I didn't know you were hurting this bad."

"Things are different now. If I had told you—"

If Margaret had told her, Caroline would have had to adhere to protocol, informing the clinicians, and Margaret would have been put on nonstop watch and under relentless assessment. "Dude, I wanted to be dead. I wanted to be gone." Confiding in Caroline would have made no sense whatsoever.

That night, new voices loomed in her bedroom. Always, her bed at night was the worst place and time to be besieged. And the arrival of new voices was a reminder that they might be numberless, that she could never count or estimate all she was up against. Her roller derby teammates knew nothing about them. By this point in her life, she mentioned them to no one. She was, with her voices, entirely alone, and that night, after Margaret had made clear why she could not be honest, why Caroline's betrayal was built into who Caroline had become, she was assaulted by a familiar cacophony urging her to punish and put an end to her failures, to take her life, while, above this anarchic chorus, a new presence predicted repeatedly that she would soon retreat forever to her parents' home, and another newcomer, a woman, in a tone empty of all accusation, said, "This can be over. We can be at peace. It's the only way. You can sleep."

The streets around the old sanitorium were filled with Asheville's hipsters; it was an easy neighborhood for buying drugs. Week after week, she kept herself addled—yet just sober enough to push through her shifts at work. She sent an email to her derby teammates, saying she had relapsed, that she needed to put all her strength into beating this, that she had to pull herself from the roster. Though they didn't know about her voices, they knew about her past with narcotics. That narrative had felt normal enough to tell.

In reply, they emailed and called: "We love you" . . . "We respect your decision" . . . "We support you" . . . "We're here for you" . . . "Whatever you need." But they had no idea what she needed. Nor did she. They sent an edible arrangement: daisies made of pineapple and cattails of marshmallow and chocolate.

"This can be over. This can be over." The woman's voice was so unlike most of the voices she knew; it didn't condemn or wail or point out those who wished her harm or insist that everyone did or instruct her to act before anyone else could. It was defined by surrender, asked nothing except acquiescence.

She kept buying drugs and quit paying bills. Her electricity was turned off. She was struck by an inspiration: to nourish and bolster herself with chicken wings. She bought a jumbo bag at a supermarket but managed to cook only a few. It was June, and the rest rotted in the warm air of the refrigerator. Maggots appeared, mounds of tubular larvae. They feasted on the wings and defeated any resolve she had left. She shut the refrigerator door, leaving them to their bloating.

"We can be together," the woman said, and Caroline set a date. July Fourth was near; it seemed as good an occasion as any. Weeks after that would be her thirtieth birthday; she would spare herself that milestone. She bought plenty of Vicodin from a neighbor in her building. She had some leftover prescription sedatives and Seroquel to add to the cocktail. She would take the pills with gin to help herself pass out. She bought extra-strength garbage bags.

The schedule at the hospital had her working on the Fourth. She was still showing up for most of her shifts and decided that this day would be no different. That evening, she would drive home, and while people were watching fireworks, she would swallow the drugs and the gin, and pull a bag over her head and bind it around her neck.

At the hospital, toward the end of the day, she locked herself in a bathroom as a co-worker was about to leave after his shift. He had an odd position in the hospital hierarchy, somewhere between peer specialist and social worker, a sort of paraprofessional troubleshooter. He and Caroline were friends; they'd traded harsh critiques of the world they worked in, and he'd told her recently about a job he thought she was perfect for. Now she hid in the bathroom to avoid any final interaction. She didn't want to be derailed by some last moment of bonding, and she couldn't bear the idea that he might think, tomorrow, that he should have seen something, heard something, in her goodbye. A small window above the toilet faced in the same direction as the hospital's front door. She stood on the toilet and waited, watching to be sure he left the building.

In a cramped office, she entered her last notes for the day. A resident knocked. Derek was tall and in his early twenties, square-jawed and broad-shouldered, blue-eyed. He looked to her like a college quarterback. He and Caroline liked to watch *Project Runway* together on a day-room TV. She didn't know exactly what had landed him on the ward. That was something to be said for peer support at this hospital: she wasn't expected to read the residents' records or listen to briefings from staff. She could avoid the diagnoses and start from scratch. She knew only what he'd told her or shared in group, that he loved rave music, that he was gay, that drug abuse was one part of why he was there. Beyond that, she knew him as boisterous and anxious to entertain, as if, having failed to be the straight, solidly built, handsome athlete everyone in Asheville would have admired, he would be ceaselessly amusing.

She turned from her computer screen. He sat down in a plastic chair. He wore a tank top and jeans shorts. He made no effort at entertainment. "Caroline," he said woodenly, before losing all composure. He crumpled. He seemed to shrink himself to fit the flimsy chair; he looked half the size he'd been a moment earlier. He told her that one year ago, on July Fourth, he'd gone out with a man he'd met on the internet. He told her that he'd been raped. He said that tonight he wanted to see the fireworks, that he needed to see them, but that the staff said they were undermanned on the holiday and couldn't organize the outing.

She drove a white twelve-seater van and pulled over near the carport. Derek and a few other residents ducked in. She knew a CVS parking lot that would give them a good view and was close enough so they could get back without too much disruption to med schedules and bedtime. Someone loaded in folding chairs. The CVS was on a hill. The residents opened the folding chairs on a strip of grass that bordered the parking lot. She and Derek scrambled onto the roof of the van.

The roof was only eight or ten feet off the ground, but the height felt luxurious. "I love being up here with you," Derek said. Below, children played in the parking lot, waiting for dark, and when dark came, the opening fireworks slithered into the sky, looking like a school of frolicking, intertwining, luminous fish. The night was the sea, and she and Derek, elevated, were in the night sea with them. Red-tailed eels came next. They rose in tight formations before jetting sideways. The eels were followed by a throng of anemones in pink and pale green, with cobalt-blue minnows slipping between their tentacles. Simultaneously, hundreds more—fish and eels and anemones and a passel of lavender jellyfish—soared and darted, writhed and danced and dove. Caroline and Derek floated. A waterfall of light cascaded upward through the ocean.

She struggled to describe for Derek what she felt. She thanked him for making this happen. "I wasn't even planning to watch the fireworks tonight," she said.

EIGHT

I asked Goff about Caroline's brain. We had been talking about the hippocampus and its relationship with the prefrontal cortex. He had been teaching me about the differences—some hypothetical, some tenuously substantiated—between the brains of those with psychosis and the brains of those who inhabit a more widely shared reality. Our conversation had ventured, as it often did, into questions of how our circuitry becomes our consciousness. For my lesson, he had cued up images of the hippocampus on his computer. He was going to lead me into the intricacies. This was the context of my question about Caroline. I wasn't asking for diagnosis; I was interested in his perspective on possible physiological distinctions. I recounted some of her story, acknowledging that my summary wasn't nearly enough information and that his response would be inherently speculative, and then asked for his impressionistic sense. "When you think about her brain, what are you seeing?"

But his reply, in his invariably quiet and compassionate voice, rerouted our conversation. "Has she been tried on clozapine?" It

seemed that he hadn't heard me. His mind had leapt to medication, to a particular antipsychotic, the first of the second-generation
drugs, the one that appeared to have the best odds of diminishing
psychotic symptoms but that also, beyond the side effects of weight
gain and permanent disorders of movement, dangerously depleted
the immune system. His mind went not to her brain; rather, it felt
to me, he went straight to the attempt to purge or contain.

Has she been tried on—the phrase sounded tangled. It was telling. I felt that he was well outside Caroline's brain and mind, that
his instinct was to draw a line around them, to bound them, to
subdue their aberrations, to prevent their anarchy from spreading.
Admittedly, this was merely my own intuition, yet *has she been tried
on* was certainly a striking construction, distinguishable from *has
she tried*. She was the object, not the subject, of the sentence, the
recipient, not the one deciding.

I said that although she'd taken a number of antipsychotic
cocktails, she'd never taken clozapine. I said, too, that although
her voices persisted, I doubted she would choose to take anything
at this point.

He emphasized that the early onset of her psychosis—in childhood as opposed to late adolescence or early adulthood—meant
that her case was probably especially severe. Then he returned to
clozapine. His language shifted. He did make her the grammatical
subject now. But he didn't seem to fully take in what I'd said about
her not being interested in resuming medication. He evinced no
curiosity about that. Was she coping? How well or poorly? He
didn't pause to ask how this severe case was doing without antipsychotics. "Someone like her will not know the potential benefit
of clozapine unless she tries it," he said. "My recommendation is
often to try it for three months. It does have some side effects. But
only then can you really make a decision about what life might be,
free of the voices and free of the paranoia."

This all sounded reasonable. Its logic was unimpeachable. It gained additional authority because Goff was a humble man. Yet it didn't feel reasonable. It skipped the question of what life might currently be, for Caroline, free of medication. It minimized and hurried past the side effects—*some* side effects?—though clozapine's were staggering even by the standards of antipsychotics. It was spoken with the presumption, the principle, that medication was the correct and best way to go.

I was tempted to postpone the images he had waiting and to press him instead about this principle. But I wanted to better understand what brain scanning suggests about the mechanisms by which we learn—and to better understand the way psychosis and learning interact. So we switched our focus from Caroline to his screen.

He displayed the graphics from a 2020 study that his Australian colleague, Esther Blessing, had led and that he'd co-authored. Subjects—half of them undergoing their first episode of psychosis and half of them controls—had lain back on the bed of a magnetic resonance imaging scanner. The machine's readings led to the brightly colored graphics Goff showed me. These images reflected what neuroscientists term functional connectivity between brain regions. Goff made it clear that the vivid hues did not represent actual, detectable communication traveling along the neural circuits between one area and another. Our technology isn't capable of that. What it can detect is the wavelike, repetitive variations in the use of oxygen *within* a region. From this data, neuroscientists proceed by inference. Their theory is that if the wave patterns within two separate regions show synchronization, then the two areas must be tightly interrelated. This is functional connectivity.

Goff and, in other conversations, Blessing were adamant about our technological constraints when it came to seeing inside the brain. What the public—and researchers—like to think of as seeing was often closer to surmising. Goff was fond of reminding me that everything he told me could turn out to be "completely wrong." And Blessing, when we sat in front of her computer, all but spurned the graphics in their 2020 paper, which had been published in a leading schizophrenia journal. The arresting yellows, orange-reds, and blues in the pictures of the brain verged on deceptive, giving the impression of more dramatic and definitive insight than science could produce. She sounded almost outraged by her own tinting, by the propensity of neuroscientists to provide such Crayola colorations, by the eagerness of science journalists to print these hues. "Voxel by voxel," she said, "you're measuring fluctuation in oxygen usage—you're getting numerical data. There are no colors, and you're not seeing a light show. You do a post-hoc analysis with very eggheaded math. And then you color." In her description, it was as if the provisional, minute advances of science were transposed into the unambiguous forms and bold palette that fill a child's coloring book.

Blessing was bothered by more than this method of presentation. She wondered if, in her field, the idea of connectivity was something of a fad. "Connectivity has gotten a *lot* of buzz and funding," she said, referring to psychiatric research that ranged beyond psychosis, "but its meaning is unclear. It's very correlative." Levels of synchronization might correspond with psychiatric disorders, but that didn't mean they played a causal part. What seemed central in the research of the moment could prove to be a sideshow.

For all their self-criticisms and reservations, though, Goff and Blessing were excited by the study and its data. "Esther is a pioneer," Goff said, "in the connectivity of the hippocampus with other regions of the brain." Blessing had explored the paired oxygen use between a portion of the hippocampus and areas that were likely trouble spots in psychosis. Her readings demonstrated that in the group of first-episode psychosis patients, who had never been medicated, the level of connectivity was lower than in the cohort of controls. Her finding added specificity, and thus a hint of concreteness, to an overarching theory that Goff and other neuroscientists had held since the 1990s. The idea was that while the prefrontal cortex—the part of the brain whose size and advanced evolution does the most to distinguish us from our primate ancestors and from all the creatures who came before, the part that houses what we term our executive function—is the ultimate locale for memory, learning, and comprehending the world around us and the world within, nevertheless the hippocampus is an essential driver of memory and learning and comprehension; the idea was that the hippocampus plays a crucial role in human consciousness; the idea was that differences involving the hippocampus are almost surely one source of the alternate realities of psychosis.

This thinking had roots, fifty years old, in the Nobel-winning research of Eric Kandel on the neural systems of learning in sea slugs. Slugs don't have hippocampi—or anything identifiable as brains—but the implications of Kandel's work spanned eons. Hippocampi appeared in fish hundreds of millions of years ago. "The hippocampus is one of the oldest parts of the brain in terms of evolution," Goff said. "It is the original mechanism by which all animals make sense of the world. Humans have added elaborate parts of the cortex on to it, but when I recognize that that's a chair, that that's Daniel, the hippocampus is at work. Recognizing patterns and recognizing that something doesn't fit a pattern, that an expectation was incorrect—what we call prediction error—this is the hippocampus."

Our circuits of attention and vigilance, reward and fear, so much emanated from this region. Within the hippocampus, through the swift adjusting and strengthening of links between the filaments that project from the nerve cell bodies—Dostoyevsky's little tails—memories are laid down. Then, sometime later, through signaling between the hippocampus and the prefrontal cortex, memories take on degrees of permanence. Another aspect of learning that relies on the hippocampus happens in the face of novel stimuli and prediction error: the reformulation of our be-liefs, of our schema, to borrow another term from neuroscience, our internal construction of the world. Listening to Goff and his expansive rendering of how the hippocampus affects our accumu-lation of knowledge—and affects our judgments about and inter-actions with all that surrounds us—I had the feeling that he was describing the origin of subjectivity itself. The feeling passed; I knew better than to be seduced by such precision; and Goff him-self added uncertainty and confusion. "I can drive you crazy," he said. "I'm always thinking about pro and con arguments, so let me give you an argument against the importance of the hippocampus in psychosis."

After laying out this counterlogic, he got back to the data be-hind his ideas. Besides Blessing's scans, he and his colleague Lila Divachi were developing evidence that in patients with psychosis, a section of the hippocampus, separate from Blessing's, responded with insufficient activity to prediction error, the mismatch of ex-pectation and experience. This engine of learning seemed to be impaired. Meanwhile, another collaborator, Daphne Holt, a psy-chiatry professor at Harvard, had run an experiment looking at the learning and unlearning of fear in the psychotic. On day one, patients and controls were taught to associate a banal photograph with an "aversive tactile stimulus," a mild shock. Later that same day, all the subjects were shown, repeatedly, the same picture without the shock, and all learned to view the photograph without fear, as measured in the palm of the hand, where fear stirs sweat and other reactions. But on day two, the patients and controls diverged. Now, viewing the photograph without the shock, the controls reacted with indifference. They had retained what they'd learned at the end of the day before. But for the patients, fear re-turned. Overnight, they'd lost what they'd learned, and a plausible reason was compromised interaction between the hippocampus and the prefrontal cortex.

For Goff, a host of published and ongoing explorations high-lighted the mechanics of learning as a culprit in psychosis; if you couldn't adequately incorporate new inputs from the world, if your schema were too rigid, you were going to be vulnerable to giving excess weight to aberrant perceptions, and your minor fears—fears that others could readily reassess and brush aside—were more likely to become full-blown paranoias.

Memory and learning had long been obsessions for Goff, and recently it seemed that his work had led to an innovative treatment. His quest for a medical breakthrough had begun, in the early '90s, with an effort to pull the field of psychosis research beyond its focus on dopamine. "For decades, dopamine was the nail and we just kept hammering away at it," Joe Coyle, a former professor of psychiatry and neuroscience at Harvard, an early mentor of Goff's and later a collaborator, told me. "If you look at the treatments we have right now, in terms of the fundamental mechanisms, they're no different than they were almost seventy years ago with the discovery of chlorpromazine." He used the generic name for Thorazine. "That's pretty scary."

Coyle remembered what it had been like when he, Goff, and others discerned the first glimmers of a new method. "You have this aha experience. It's kind of narcissistic; you know something that is consequential. It's like you've figured out how God designed this one small aspect of the universe." The one small aspect was a receptor, a molecule that sits at the tips of the tails or on the surface of the nerve cell body, and receives signals from particular neurotransmitters. The molecule that had thrilled Coyle and transfixed Goff is abbreviated as the NMDA receptor, and it helps to regulate a range of excitatory and inhibitory—spurring and slowing—processes in the brain. Goff led the attempt to employ it in medication. "I was just tagging along," Coyle said. "Don Goff is the best clinical psychopharmacologist that I know"—the best at translating data garnered in the lab, often starting with rodent experiments, into means to help human beings. Goff had a chemical compound, DCS for short, that bound to a site on NMDA receptors and, in a preliminary study published in 1995, helped patients with psychosis not only in tests of memory but also in social interactions. This brought funding from the NIMH for a larger, placebo-controlled study, which replicated some—though not all—of the first trial's success. And this led to a yet larger, more rigorous trial, completed in 2007. Goff's compound, DCS, offered patients no benefit at all. "I would have sliced my wrists," Coyle said.

Goff did not give up. "We ran into disappointments," he said, "but we always had a hypothesis for why and a new idea for how to make DCS work." By the time he and I met in 2019, he had published twenty peer-reviewed papers on DCS, and two sentences from the introduction of a 2012 paper allude to the length and detours of his journey: "Roughly twenty years ago, we considered DCS the best available agent to test the emerging NMDA hypofunction model. . . . Our understanding of DCS activity at NMDA receptors has changed significantly since that time, leading to novel therapeutic strategies that recently have shown promise in early trials."

The most hopeful of these newer strategies built on cognitive behavioral therapy, CBT, a type of counseling that tries systematically to change destructive habits of thought, an approach that has shown some—but not great—effect in psychosis. Because of the evidence that DCS can enhance memory and therefore might help to consolidate learning, Goff's theory was that it could make the lessons of CBT take stronger hold. Might it help to free psychotic patients of paranoias? Rid them of other derangements?

Initial studies delivered suggestions that Goff was onto something. There were complexities and contradictions in the data, as, with the brain, there always are, but the suggestions were enough to nurture Goff's belief—belief that had grown over decades of creative scientific thinking, of tedious grant writing, of arduous managing of experiments, of interpreting outcomes with an acknowledgment of the imperfections yet an emphasis on the subset of measures that more or less confirmed his theory. He won funding for a larger trial.

And the trial was a failure. In the paper that followed, he was blunt: "Conclusions: DCS augmentation of CBT did not improve delusions compared to placebo during treatment." These straightforward words were followed by a semicolon and a "however." He appended a reference to a portion of data that supported continued faith. But it was a pinky hold on a towering rock face; he was exhausted; he was letting go. "A clinical trial can take five years to complete, and before that, in order to get funded, you have to write a really compelling proposal. And in the process of doing the writing, you convince yourself that this is perhaps a great scientific advance. You're so optimistic, even though experience should teach you otherwise. And then, after the five-year wait for results, if they're negative, you think, did the placebo somehow get mixed up with the drug? Did we somehow get a defective version of the drug? I've essentially abandoned DCS research. It's heartbreaking."

I asked what was next.

He sighed, a long exhalation, then rallied. He could always rally. He spoke about revelations latent within the hippocampus and its circuitry, and about pathways running through the thalamus, a switchboard at the brain's center. He mentioned a type of ultrasound that one day might—by stimulating or inhibiting neurons—prove useful to understanding and treatment. But he said, "The big picture is that we really know very little about psychosis and what is going wrong in the brain. It's remarkable what we don't know."

By this time, I knew the list too well:

In readings like Blessing's of the brains of psychotic patients and controls, the distinctions held up only in averages. The over-lap between individual patients and individual controls was such that the data could not be used diagnostically. A healthy subject could have a pattern of connectivity that looked like it belonged to someone with the condition, and vice versa. Science, with the technology currently at its command, cannot say, This brain be-longs to someone with psychosis.

And it cannot even begin to say how the hallucinations and delusions of psychosis are implanted in the first place. What, in physiological terms, were the voices Caroline heard coming at her from outside her own head? They didn't consist of sound waves. Yet they were heard just as if they did. How did these phantasms arise? How did they insert themselves into her mind? Science had no clue.

Nor did it have much more than nascent clues about the main causes of the condition. Or rather, there were so many clues as to leave science lost. As I talked with Goff about the demise of DCS, I had just reconnected with Steven Hyman, a past director of the NIMH and now the director of a joint Massachusetts Institute of Technology and Harvard center delving into the genetic under-pinnings of psychiatric illnesses. But I already knew, from talking with Hyman's team, that psychosis could be tied to around three hundred genetic variations, as compared with a neurological dis-ease like Huntington's, which is due to a variation in a single gene. And the count of three hundred was growing rapidly. It all but ridiculed genetic research.

Studies suggested that a person's environment, as early as the womb, was full of risk: Maternal infection or poor diet or stress could raise the odds that a child would eventually develop the condition. Childhood trauma posed a risk that was probably heavy but was hard to quantify. Growing up in a city seemed to be a danger, perhaps because of factors in the social fabric or be-cause of pollution or higher exposure to bacteria and viruses—

immunological reactions, whether in the expectant mother or later in the child, came up often in the epidemiological research.

Perplexingly, well-replicated findings showed that birth season was a factor. Being born in winter or spring was associated with higher rates of the disorder.

And there was alarming data from British scientists showing that black Caribbean immigrants to the UK had a sharply increased risk of psychosis. This was possibly due to bias in diagnosis, with practitioners being much quicker to detect frightening mental illness in patients of color, or due to greater use of cannabis, though this explanation was fiercely contested. It seemed, finally, that the disorder might be triggered by the battering and alienation of bigotry and discrimination—and that the stress of poverty might contribute. Prevalence was heightened by nine times in the UK's black Caribbean population.

All in all, potential causes of the condition were myriad, their relative importance a muddle. And this was a hint that the disorder might actually be dozens of disorders, deemed similar only because of our need for categories, for containers, a need perhaps especially powerful with psychosis, because its alternate realities can be so unsettling to those of us who do not perceive them that we long to bundle them together, as if to more easily push them out of view.

Still, science proceeds almost blindly, from scattered intimations, with a faith that the brain will gradually reveal its secrets, though its principal secret may not be concealed at all, if it is that the brain is more than the sum of its parts, that the brain and the mind are related but vastly and ineffably different. This would help to explain why Kennedy's ambition of reaching the moon was so swiftly accomplished while his rhetoric about making "the remote reaches of the mind accessible" has proved delusional. The moon is there, physically, to be reached. With the mind, this may not be so.

)

Angelica Torres-Berrio prepared the rodent lab for a test of resilience against stress, a step in the search for new antidepressants. Torres-Berrio, a young neuroscientist on Nestler's team, had grown up amid the five-decade civil war in Colombia. Revolutionaries had forced her father to flee his village as a child. The question of why some people wound up devastated by stress and trauma, while others came through fairly well if not unscathed, had drawn her to Nestler's work. This was the question at the core of his research, ever since he'd heard his colleague lecture on the durability, the psychological armor, of some POWs in Vietnam. That lecture had been seventeen years ago. And just now, Nestler had the results from a trial of a new medication that had emerged from his labs.

Work wasn't stopping because results on one molecule had come in. Torres-Berrio wheeled a cart of mice, dozens of them in stacked cages, into the windowless room that was the size of a walk-in closet. She set up four boxes, so she could run four mice simultaneously through her experiment. If she went one at a time, she would be in this room, a kind of cage in itself, from first thing in the morning until well into the night. As it was, she would be here half the day. The musky ammoniac smell of mouse urine was thick in the air.

Each box, one and a half feet square and open on top, contained a mini booth at one end and zones mapped out on the box floor, one zone near the booth and others in the farthest corners. She adjusted overhead cameras that would monitor movement in the boxes. She dimmed the lights to a level that the mice wouldn't mind and reattached a sound buffer that was drooping away from a wall. By their tails, she plucked four black mice out of their cages. She dropped them into the boxes, then measured where they liked to spend their time and how much they liked to explore. Next, she placed a white mouse, a much bigger and more aggressive breed, into each booth of crisscrossing wire.

Before all this, most of the petite black mice had been systematically terrorized. In childhood, for five minutes each day, for ten straight days, they were penned up with a hulking bully. They were bitten, chased, lacerated, ambushed, mauled. When the five minutes were over, the ordeal wasn't. After these sessions of torture, the child mouse was left overnight in the same cage with the brute, protected by a solid but transparent screen, so the trauma continued around the clock. Known as the protocol for chronic social defeat stress, it was one of a number of methods for inflicting trauma on lab rodents. A newborn might be separated for long stretches from its mother. Or separated as well as deprived of soft bedding. Or a mouse might be left dangling upside down by its tail. This would be done at varying times, so the mouse never knew when torment was coming and constantly felt it was near.

Now Torres-Berrio—with the help of the cameras and a pair of computers, so that human fallibility didn't distort the measurements—determined which of the black mice had endured chronic social defeat without much damage to their exploratory and interactive instincts and which had been psychologically marred. This only became clear once the belligerent and bulky white mouse was in the wire booth. The white mouse couldn't get out, but it was still an intimidating presence. Before Torres-Berrio set it in the booth, all the black mice did a good deal of poking, sniffing, and scooting around the box, despite their history of trauma, because poking and sniffing and scooting is what mice do in a new environment. They want to learn about their world. They're driven to gather data. Call it a survival instinct or call it desire; it amounts to the same thing: it's an essential part of a mouse's being. But when Torres-Berrio lowered the giant into the booth, differentiation occurred. Some mice—perhaps forty percent—went right on poking and sniffing all over the place. More than that, they went up to the booth and nosed at the wire bars. They stood on their hind legs, as if to check out the prisoner from another angle. They nose-kissed the big white mouse. They scampered away, lingered in the middle of the box, twitched their heads, and returned to the booth for another nose kiss. These were the resilient ones. Regardless of their traumatic pasts, their basic beings were intact. Just like a cohort of control subjects who'd been spared the ten days of mauling, the resilient mice were avid about seeking both information and social activity.

The rest, the mice who were more susceptible to the ten days of stress, hunched in the corners of the box farthest from the booth. There they swiped their paws repeatedly, at hyperspeed, over the sides of their heads, a gesture of grooming that reflected high anxiety. They stood on their hind legs but only to examine the seams where the box walls met, as if they wished to escape. They approached the mouse in the booth for an instant before quick-stepping back to a distant corner.

Soon the resilient and susceptible would be euthanized, and, in another of Nestler's labs, a large room brightened by a bank of windows and busy with researchers hovering over brain-slicing machines and peering into glass tubes containing ivory-colored beads of mouse RNA, Torres-Berrio would freeze the brains of her subjects and cut them into wafer-thin sections using one of the machines or an elegant manual instrument that looked like a jeweler's tool. Employing a minuscule version of a cookie cutter, she would punch out the brain region she wished to study, collect the punched-out tissue of the resilient and susceptible cohorts into two batches, pulverize and turn each batch into soups, and analyze their chemistry.

This was basically how Nestler and his team had arrived at the antidepressant whose efficacy had just been put to a placebo-controlled test. Mice had been subjected to chronic social defeat stress. The resilient had been designated not only by their enduring tendency to explore and interact, but also by their interest in sipping sugary liquid over plain water, their ability to maintain weight, their steady patterns of sleep, and their relative indifference to getting high on cocaine, in contrast to their susceptible counterparts, who seemed to take no pleasure in the sweet drink, mirroring a depressed symptom in humans called anhedonia, who lost weight, who had trouble sleeping, and who appeared, judging by a Skinnerian test that associated place with altered state of mind, to crave more of the narcotic.

In the susceptible mice, Nestler had found heightened and lowered quantities of certain proteins in a set of brain regions, and in an autopsy study of human brains, he'd found analogous protein levels in the same regions in the depressed. What fascinated him even more, though, was the adaptive ability he discovered in the brains of resilient mice: the activation of molecular chains of events that seemed to counteract the problematic protein levels. Then, in a process that took eight years, he'd found that a seldom-used epilepsy medication induced a reaction that mimicked one of the adaptations of resilient brains. An injection of the medicine made susceptible mice behave like resilient ones. And in a preliminary study, with eighteen depressed patients and no placebo controls, the drug, ezogabine, relieved symptoms to a notable degree.

Yet just as funding perhaps should have poured in to develop this potential antidepressant by running trials of increasing size and rigor, progress had stalled. Funding was scarce. The pharmaceutical companies weren't all that interested. To Nestler, this was infuriating but not quite surprising. He'd run into two well-known obstacles.

The first dated back to early in his career, to the late 1980s and '90s. He'd tried to persuade the companies to think beyond their conventional approach to medicating depression, which focused on serotonin and which dominated the industry for decades. He tried to convince them that their methods were too crude. Though he didn't speak to the industry's research departments so derisively, he believed that the prevailing technique was the equivalent of treating the brain like a water balloon and pumping serotonin into the water. Somehow the drugs had some antidepressant effect for some people, and meanwhile, because they were woefully untargeted, they caused a panoply of side effects, from weight gain to impotence. In presentations to the industry's pharmacologists, he showed slides detailing processes inside the neurons, as opposed to the chemistry in the clefts between the cells. It was by searching *inside*, he argued, that science stood a chance of finding the more primary mechanisms of psychiatric disorders, of locating targets so that medications could have much greater efficacy and inflict much less collateral damage. The companies were unmoved. Prozac and its spinoffs were new to the market, supplanting the tricyclics, replacing one set of water-balloon medications with a better set, superior on one count, that they didn't have a dangerous cardiac side effect and so could be dispensed widely and without concern. Billions and billions were being made. And up against such profits and the entrenched synaptic theory, Nestler made little headway.

The second obstacle had solidified more recently. It hardened even as Nestler's ideas about internal mechanics gained traction and appeared in textbooks, and as serotonin elicited more and more eyerolls among neuroscientists. This second barrier was that, in the new millennium, the pharmaceutical industry drastically cut its investment in finding new psychiatric drugs—by as much as seventy percent between 2006 and 2016. The sector had confronted its fundamental lack of progress. It had lost faith in exploration. "From the point of view of the shareholder," Nestler said, "the cutback makes perfect sense. Better to invest research dollars in cancer, in diabetes, in infectious disease."

But Nestler had managed to conduct the placebo-controlled study of ezogabine, a small trial with forty-five patients. In the medication group, sixty-one percent of the subjects had felt their depressive symptoms diminish by half or more, compared with thirty-seven percent in the placebo cohort. Telling me about this, he sounded almost exultant. The NIMH had just agreed to fund a larger trial. That would take two or three years, he said, and if those results were good, a third phase would follow, with many more hundreds of patients and more years of waiting. For a moment, the future tests didn't seem to worry him. Yet the current results contained warnings. His theory about how ezogabine worked wasn't backed up by scans taken during the study. Might this theoretical issue point to an arbitrariness, to sheer good luck, lurking in the study's outcomes? Would the gap between medication and placebo shrink toward negligible in the upcoming trials? And, too, there had been some side effects: dizziness, drowsiness, a strange discoloration around the lips. There was no telling what would come with long-term use.

Nestler and I turned to broader topics, to our society's over-reliance, he said, on psychotropic drugs, to our reflexive seeking of medication to salve our emotions, and to the psychiatric profession's reflexive writing of prescriptions. "Unquestionably," he said, about whether fewer Americans should be on psychotropics for depression and anxiety. "We could do with much less medication."

There were caveats. He clarified that medication was a necessary part of treatment in severe cases. And he worried that in communities of color some might not have access to psychiatric help, or might not accept medication because they did not trust practitioners.

Then he went on: "Exercise. Better sleep. Mindfulness. The belief in something bigger than yourself. Religion if you're religious." These were antidotes, inoculations, boosters of resilience. The last item on his list caught me by surprise. Religion wasn't something I heard much about from neuroscientists. "People with religious beliefs benefit greatly from them."

I asked if religion was part of his life. It wasn't. "The thing about religion is, I can't know whether Jesus is the Son of God or whether Allah rose to heaven on a winged horse. Those are not scientifically knowable." This, for him, was an impassable deterrent. "But I like to think about the active ingredient in religion, or in mindfulness, that gets a person to a better state. Perhaps to a metaphysical level of comfort. It may be the capacity to bring order and meaning into one's life."

From metaphysics, it wasn't much of a leap to talk again about consciousness. "My son and daughter-in-law have a dog, and it's the first dog I've ever gotten to know in my life. I love this thing. And as a neuroscientist, I spend a lot of time wondering about what she's thinking. It's impenetrable. Does she think, *I'm going to die in ten years, and that will be the end of me*? Who knows? Who the fuck knows?" He segued from this existential question to a problem of "what philosophers call emergent functions—functions that arise as a whole that are not evident in the molecular and cellular constituents of the organ. This occurs only in the brain. In every other organ in the body, the function of the cells matches the function of the organ. Heart cells pump blood. Lung cells add oxygen and remove carbon dioxide. Kidney cells filter the blood. I can cut out a chunk of a kidney, put it in a petri dish, and show you that it does just what it does in the animal. I can show you a single heart cell pumping. The brain is the only organ where this is not so. And that's what makes it really cool! Brain science is unique in medicine. We need to understand how circuits of cells give rise to a thought, an emotion, a behavior. And this will be extremely difficult to penetrate."

I heard an echo of what he'd said about the consciousness of his beloved dog; the impenetrability might be on the same order. I asked if it was even possible that we could solve this dilemma about ourselves. I asked if it was even possible to *imagine* crossing the chasm between the brain and the mind, between the activity of cells and our thoughts and feelings. "How do we bridge that divide?"

"That is the crux," he said. "If I knew how to cross it, I'd be a very special person."

WHEN STEVEN HYMAN thought about the brain, he thought about Borges. He thought about Theseus. As a child, in his elementary school and junior high classrooms, he'd lost track of whatever he was supposed to be doing; he was too busy drawing labyrinths with the Minotaur, the man-eating monster of Crete, waiting at the dark center. At the Broad Institute of MIT and Harvard, Hyman ran the country's largest effort to parse the genetic components of psychiatric conditions. The institute's project dated back to 2007—he had taken over in 2012—and amounted to amassing the genomes, the entire genetic scripts, of hundreds of thousands of people, some diagnosed with disorders and others not, and then trying to sort out which genetic variations were associated with which conditions.

Hyman's work did not mean that he had taken a side in the eternal nature-versus-nurture debate. Even in a physical characteristic such as height, both nature and nurture play their parts, and in psychiatry, nature and nurture each contribute numberless elements. Among identical twins, if one twin develops the hallucinations and delusions of schizophrenia, then the other twin will have the disorder in forty to fifty percent of cases. But outside of the rare demographic of monozygotic twins predisposed to schizophrenia, genetic impact gets much weaker. Having one schizophrenic parent will increase your chances of having the condition from just under one percent—the prevalence in the general population—to around six percent. Suddenly genetics looks important but a lot less determinative. And with most psychiatric diagnoses, from depression to obsessive-compulsive disorder to ADHD, the lines of genetic influence are yet more tenuous.

Hyman and his colleagues were using fast-evolving genomic

technology, advances he liked to compare, for their revolutionary power, to the technological breakthroughs in telescopes that
allowed Galileo to rearrange humanity's understanding of the
earth's relationship to the sun. Hyman's team was isolating relevant genetic aberrations that could then be tied to mechanisms
within the brain. The aberrations would let his scientists pinpoint
problems in cells and circuits. The technology was, for him, something of a miracle, and its sorting of genes seemed to supplant
deductions from mouse behavior, which he called "anthropomorphic
fantasies," and the imperfect insights from brain scans that dominated much of neuroscience research.

Yet when he'd embraced the genetics of neuroscience, he was,
he said, "terribly naïve. Have you read Borges's 'The Garden of
Forking Paths'? The brain is a garden of billions of forking paths.
And as I got deeper and deeper, I realized that I had so underesti-
mated the magnitude of the complexity. If you'd told me that there
were hundreds of genetic variants creating little nudges inside the
brain toward schizophrenia or depression, I wouldn't have believed it, but that's the way it is. Nature is not kind to scientists."

One of Hyman's collaborators, Benjamin Neale, a statistical
geneticist, added that the ever-growing tallies of pertinent variants with tiny, subtle effects would likely prove to be the case
across psychiatry. "The only reason we haven't hit two hundred
and fifty in ADHD is that we haven't found them yet." And the
hundreds held mind-boggling permutations. "And then," Neale
said, about the range of psychiatric disorders, "there are all the
environmental influences, starting with what your mom ate when
you were in utero and ending with who you just had lunch with."

Hyman thought back to the beginning of his career, when the
biological understanding of disorders—serotonin deficiency for
depression, dopamine excess for psychosis—was shaped by the
blunt drugs that treated them. "Psychiatry has lacked a ground
truth. It's a house of cards built on serendipitously discovered

drugs. How people could think that mediocre—important but mediocre—drugs like the SSRIs could give us any comprehension is beyond me. The logic is like saying, I have pain so I must have an aspirin deficiency."

He spoke about the need for "epistemological humility." He reckoned not only with the dizzying numbers of relevant variants and their combinations but with a misleading formulation put forward in 1980 by the DSM-III. In their desperation to establish psychiatry as a medical science, the framers had promoted a way of thinking about diagnosis that was categorical, definitive. A patient either did or didn't meet the checklist threshold for a disorder. Diagnostic boundaries were much overemphasized; the idea that we might all exist along psychological spectrums was pushed aside. This still had a deceptive resonance that hindered the quest for understanding.

Yet the actual "fuzzball" quality, as Hyman put it, of disorders, the lack of sharp delineations, the infinite individuality of the psyche, posed a barrier to genetic research. Genetic patterns were worthless unless they could be attached to diagnostic categories. He needed the boundaries he derided. It was a paradox that brought him back to his memories of sitting in class and drawing, with number 2 pencils, pictures of the subterranean labyrinth and the Minotaur, the half-man and half-bull who feasted on human flesh. In the Greek myth, Theseus, the son of the Athenian king, boldly enters the maze of tunnels and slays the monster. These days, Hyman thought of himself as wandering unlit labyrinthine passageways. Such was the brain; such was the search for knowledge. But he didn't have Theseus's sword. He didn't have the magical help of Ariadne, the princess of Crete, that lit Theseus's way. He was venturing through the labyrinth, more or less blind.

I asked how he avoided scientific nihilism.

"I'm always excited about the next result. It's just that I have to be clear-eyed about what it means hours or days later." He said, "The Minotaur hasn't killed me yet."

NINE

In his second session with Dune and psilocybin, David did not have another encounter with the blurry fetus, but he did gaze into a golden room. It had no walls; it swirled; it expanded. Its gold was radiant, its space intangible, its effect one of awe. But neither the fetus nor the room nor anything else about his psychedelic journeys seemed to have a lasting impact on David's psyche. Perhaps, he thought, his suspicion had been right: his rationalism was too ingrained, leaving him unreachable by this method. There was also a pattern in what he'd seen. The fetus was fleeting; the room was there to be viewed but not entered. He recognized, after the second session, that even in an altered state he retained the sense of being separate, cut off from something inarticulable, something that might complete or cure him. Even on a psychedelic trip, he couldn't escape himself.

There were moments, in the days after each session, when a lessening of one symptom or another, an extra forty-five minutes of sleep or a more limited attack of nerve pain, tantalized him with the prospect that Dune and his drug had done him some good, that he was turning a corner, but it was never long before he confronted the facts that his insomnia persisted and that his bouts of scorching nerves, from feet to tongue, continued mostly unabated, and that he went right on feeling that his brain consisted of cotton and that, at work, he was incompetent and uncaring.

The November election was only several weeks away. If Trump lost and refused to concede, if he negated the results and insisted on clinging to the presidency, if progressives and liberals marched in protest, would officials in David's region, officials who were Trump allies, impose curfews? And whether they did or didn't impose them, and whether they did or didn't request federal help, what would Trump see fit to do with the military he still controlled? David's organization met to take proactive steps. The attorneys discussed sending letters to officials, reminding them that their curfews during the rallies of the spring and summer against police brutality had been ruled unconstitutional and warning them not to repeat that attempted injustice. David was speechless during the Zoom meetings. Afterward, he cobbled together a letter by cutting and pasting from one his office had sent months earlier; he couldn't muster any new turn of phrase or slant of argument.

Outside of the letter, a new phrase did occur to him. It was "compromised mentally." He spoke it to himself, about himself, and imagined what would happen if a judge decided to hear emergency oral arguments on a curfew case or anything else related to the election. Of all his talents as a litigator, parrying verbally had once been the skill he was proudest of and most confident about. Now he envisioned himself disintegrating in front of a judge. He foresaw a scene of logical stumbling, verbal torpor, legal slapstick, pathos.

He checked and rechecked political websites uncountable times each day, comforted by any semblance of a prediction that November 3 and all that followed would be orderly, sparing him the humiliation of legal action—but his mind was dominated not only by his incapacity and current lack of passion but also by the insidious idea that he'd never been passionate about civil rights in the first place. The idea had snuck into his thoughts, and lately it popped up more and more. He knew, or partly knew, that this indictment of himself was irrational, and yet it was as if, in place of his lost legal acumen and ardor, his mind had marshaled a prosecutorial flair directed at his own failings. Even going back to his triumph against the prison system superintendent and his abusive guards, he found reasons to impugn his commitment.

Add to this the certainty that he was the rare public-interest lawyer in America who'd somehow been paralyzed rather than mobilized by Trump's presidency, who'd been harboring the impulse to flee Trump's horror, and he became close to irredeemable in his own judgment. Surely he was a liability to his own organization, a weak link that could be disastrous if the organization had to go into emergency mode in November. He debated telling his boss all that was going on—and that he needed to be demoted, removed from any issue of urgency, put out to pasture in some kind of avuncular, mentoring role, or that it might be time for him to leave and find another, softer line of work. He decided that to confess and cut short his career would be the moral thing to do.

There was a mesquite tree in the family's yard. An old friend of Amanda's, an arborist, had given it to them in the weeks before Gillian was born, so the child and the tree would grow up together. One late morning, he and Amanda lay back under the contortions of its branches and its stubborn shade.

The night before, he'd given her a massage, not just her feet—their usual limit these days, because he was as thwarted and frightened in this as at work—but over much of her body. The massage did not lead to sex. She had noted, other nights, that it had now been over half a year since they'd had sex, and he absorbed these comments not as a kind of invitation but as a condemnation and dismissal. Yet last night she had told him, instead, how nice it was to be touched, how much it meant to her. And this had led, a bit before noon the next day, to their taking a break from work and lying together on an outdoor couch under the mesquite tree.

Its trunk was multipronged and tilted, its trifurcations growing knobbily at thirty- or sixty-degree angles to the ground and bending back on themselves. They sprouted branches that seemed to be trying to form wavelike shapes or even to return to the earth. But the total effect, though the tree's foliage was feathery and almost sparse, was a uniform, graceful shade, exactly enough to screen but not eliminate the sun.

Amanda said that life was strange, and they held hands. He said that it was amazing: how much the tree had grown. They kissed. She said, as she sometimes still did, that she respected the fight he was putting up. He told her how much he appreciated her saying that. They kissed again. They left it there, and soon she was gone on the solo vacation to California that she'd been threatening to take for months for the sake of her own sanity.

He tried to keep track of signs that he would recapture his life: the half hour under the tree. A conversation he had with himself, for which Dune might deserve credit, about forgiving his mother her rigidity and distance. The words of his predawn internet friend in Iowa, to whom he'd written, asking, "Can you say anything to encourage me?" and who had replied, "You will heal. I was as bad as you are. Each hour you suffer is an hour closer to the finish line."

But was there a finish line? Were his symptoms a reaction to getting rid of the pills? Who had he been before that, and who had he been before he'd ever touched an antidepressant? In the article about withdrawal that he'd read in the *New Yorker*, the main character talked about wanting to rediscover her original self before the brain alterations of psychotropics, the chain reactions set in motion by the drugs, the mechanical and structural adjustments that could be long-lasting, changes that few researchers in the field now doubted but none could specify—she talked about the need to find herself no matter what the cost and about the pleasure and perspective of finally succeeding. But what if much of this was who he'd always been? He thought in terms of Eeyore, Pooh's ever-gloomy friend. What if Eeyore was his fundamental self?

With Amanda away, his worst fears about being left alone with Gillian were not realized. He did not collapse and fail to shop for food and abandon their daughter to scrounge for meals. They cooked her favorite Chinese dinners together. He did not fail to arrange a weekend activity or two for them to share. He did abandon a plan to go camping, because he lacked the focus to research campsites online, but he made sure to take Gillian on a walk around a neighborhood whose architecture they loved and hadn't seen for years, to coax Gillian into helping him restore a piece of furniture that he'd promised would be stripped and re-finished before Amanda returned home, and to rebound for Gillian in the driveway during extra-long sessions on her perim-eter shooting. Though they never discussed it, she seemed to know he was struggling. She seemed to forgive him, to want his com-panionship, though they often fell into unbroken silences.

But the reasons to trust in a finish line were scant. He reread *Darkness Visible,* fixating on the familiarity of Styron's thorough loss of self-esteem, his descent from award to hospitalization, and oblivious to Styron's resurfacing within the course of the slender book. David's days and nights were cluttered by a voice—not like those of psychosis, which come from without, but a voice within—asking: *How are you feeling? Any better than when you woke up? How are you feeling right now? How are you feeling right now?* It was parrot-like, mimicking itself, mocking him.

He heard about the Fischer Wallace Stimulator for depression, anxiety, and insomnia, and read up on the device. The featured research on the company's website was a study with nine subjects in the treatment group, but he paid three hundred dollars and days later followed the instructions. He loaded double-A batteries into the small white gadget, connected red and black wires, soaked the head sponges in a glass of cold water, set the band snugly around his head, wedged the wet sponges under the band and against his temples, and switched on the machine. *What are you feeling right now?* his internal inquisitor asked over the subsequent days of use, *What are you feeling right now?*, and the answer, *The same*, added to the evidence that he was beyond reclamation.

When Amanda returned from her hiatus, and he reported on his condition, she posed an impatient question, one she'd never asked before. She referred to a neighbor, a woman with multiple sclerosis. The woman maintained an upbeat attitude and spoke about self-acceptance, though she was deteriorating, her body degenerating, her mobility horribly limited.

"Why can't you accept who you are?" Amanda asked him.

"Because I don't want to accept who I am!" he cried out.

MY BROTHER LEFT the hospital feeling a blend of resistance and surrender to his diagnosis, the resistance stubborn but the surrender having the final say. One night he wandered into Jazz Alley, listened to a Latin band, saw himself at the piano, saw dances he would choreograph, and knew, absolutely knew, this was what he needed to do with his life. But no sooner had he reached this certainty than he was overtaken by psychiatry's definitive term, "bipolar," along with our mother's admonition that any thoughts of a career in music were a manifestation of his illness and meant inevitable suicide. After the night of Latin jazz, he described his renewed ambitions and confidence to our parents. The three of them agreed that he should see his psychiatrist right away to have his medication adjusted.

"Why did I believe any of that?" he asked me, asked himself, later. He talked about being under a spell cast by our parents and by the profession. "Diagnosis can ruin people's lives," he said. He was not speaking solely about medication; he was dwelling on the cage of diagnosis itself.

This was the other side of biological psychiatry's claim of removing stigma. The claim has a strong logic. Most disease is not seen as anyone's fault; it is visited upon us. To walk out of a psychiatrist's office with the name of an illness and the idea of biology beneath one's torment can be a relief or even the cause of a surprising, short-lived euphoria, and it can embolden people to talk openly about what they're confronting. Isolation can be reduced, softened. But it can also be augmented, hardened. Biological diagnoses put us on the other side of divides. They do this efficiently; we can see it on people's faces, in their immediate reactions when we use this or that terminology about ourselves or someone we love. We catch, in their eyes, sympathy but also a pulling back. People may not blame biological illness on the afflicted—though often, at some conscious or unconscious level, they do blame—yet even when there is no rational threat of contagion, there is a recoiling. People don't want to be too near the diseased.

But maybe the way biological psychiatry most stigmatizes and isolates is by alienating us from ourselves, by defining and circumscribing and sometimes damning and imprisoning us in our own eyes. It is interesting that the profession may do to its patients precisely what we are advised not to do in raising our children. We are told not to define them in sentences that begin "You are" or to have them overhear us say "She is," but rather to speak of what they do and say and the effort they invest. The first mode, even when the words are flush with praise, delineates and diminishes, while the second leaves the child feeling seen yet not delimited; it preserves her own sense of herself as an ever-evolving human being. It nurtures and protects her paradoxical belief about what the self is: both actual—*I am me!*—and highly malleable. Biological psychiatry risks doing the opposite, putting us at odds with the feeling that we are protean and full of possibilities, and teaching us, instead, to conceive of ourselves in permanent terms.

Besides artistic ambitions, Bob raised something else with our parents as a sign of oncoming trouble. He'd been told to watch out for his appetite, but he'd returned to eating his usual large portions of chicken and pasta. There was no concern that he would become portly let alone obese, though sizeable weight gain can be a side effect of lithium. Between dance class and working out, he had always gotten lots of exercise and didn't let up after the hospital. His body was chiseled without being bulky and must have been the envy of others at the gym. The concern was somehow that he might eat too much and pack on too much muscle. If this sounds bewildering, I can only say that perhaps anything uncommon, anything heightened, even strength, was viewed with suspicion.

He confessed these issues to our parents and then to his psychiatrist. Whether his dose of lithium was raised at that point is a detail lost to imperfect memory and missing records. It may not have been, if only because an increase might have been seen as dangerous and a step to be postponed, given that his dosage was already above the upper end of the conventional range.

The drug, an organically occurring salt whose full name is lithium carbonate, had a circuitous history. It had been used to treat kidney stones in the nineteenth century and, in the early twentieth, included in the lemon-lime soda 7-Up and touted as an ingredient that quelled nerves and enhanced energy. At mid-century, it was removed from the soft drink amid evidence that it could cause the kinds of problems Jamison would suffer in serious form and write about decades later in *An Unquiet Mind*. Yet psychiatrists were finding that the salt seemed both to flatten the highs and bolster the lows of manic depression, so, despite the drawbacks, despite the fact that beyond trace readings lithium carbonate isn't a natural compound in the body or brain, and despite the fact that no one had more than a vague concept of how the substance worked neurologically—a mystery to this day—the drug gained psychiatric support and was approved by the FDA in 1970.

Between resistance and surrender, Bob lived for three or four years. Then, giving a recital at his piano teacher's house, he became acutely aware of the tremor in his hands. At his prescribed dose, the tremor may have been preordained. He'd been living with it since the hospital; this was part of the surrender; but now, as he performed a Mozart sonata, as he drew from the keys a mix of gentle throb, pulsing defiance, and delicate lilt, he noticed it more than ever, and during the sonata's slow, ethereal section, he could see his fingers quivering on the keys, their involuntary movement straining toward musical chaos. He finished the piece, hearing the notes only intermittently and hearing above them a decision: *I'd rather be dead.*

This was in Washington, D.C. Our parents had moved there for new jobs, and Bob had followed them, another submission. He could have stayed in Seattle or joined me in New York, yet he chose to stay with them—not live with them, but be near them, in a room he rented outside the city. He wanted their approval of the ideas he still held about his future and of the notion that his diagnosis was mistaken and his medication unneeded, of all that he occasionally slipped into conversations and that they succinctly rejected. He wished for their affirmation. How different was he from any of us in our twenties—or fifties or sixties or until we die, hungering in some corner of our psyches for our parents' praise, even when they are long gone? He couldn't keep his distance.

Yet spurred by the anarchy of his fingers and by the feeling of having a blanket over his brain, another typical side effect, documented even by the psychiatrists who, back in the 1960s, took lithium themselves to better understand the drug, he bought gallon jugs of water and checked into a motel room. He didn't know what to expect and didn't want to wrestle with whatever would happen in the apartment he shared with strangers. He made sure the motel room door was locked and began to drink, to flush the medication from his system. He pissed multiple times an hour and panicked over what felt like the electrification of his brain and stayed in the shower with water crashing over his head for long stretches and saw flashings and floodings of light. He walked out of the motel room a day later, having done what next to no one in the psychiatric profession would have advised. It wasn't only that he'd cut off his lithium cold turkey; it was that he'd gone off the medication at all.

The intangible presence he'd felt on the Olympic Peninsula, the presence that had tracked him to Seattle, reasserted itself, unassertively. The ending of *Franny and Zooey*, whose message had so elated him, whose lines deliver to Franny a lesson about humility and art and serving others and leave her feeling, Salinger writes, "as if what little or much wisdom there is in the world were suddenly hers"—those sentences, too, reverberated. Bob quit his job as an office assistant at a college and volunteered full-time with a new Christian ministry, restoring houses for low-income families in a crack-ridden neighborhood of D.C. The ministry's founder listened to my brother's story of flight from the psychiatric system. In the red-brick Victorian house that was being renovated as the ministry's church and children's center, he gave Bob a turret room, about eight feet in diameter, where he could keep a duffel of belongings and where he could sleep.

He'd picked up some carpentry skills back in Seattle, and for a time he framed walls and hung drywall and built stairs. Then— another fragment lost, another reason impossible to reconstruct— the mission's founder called our father and called the police. He wanted my brother out of the turret. They all gathered in front of the ramshackle Victorian house on the street of dealers: my brother, our father, the pastor, the cops. Bob refused to respond to anyone except the police. He told them he would go upstairs and pack his bag and leave, if they would guarantee one thing: that our father would not touch him on the way out. He was afraid that our father would lay his hands on his shoulders and tip his head forward and gaze out of the tops of his eyes and persuade him to rehospitalize himself and return to a life of lithium. "Yes, sir," one of the policemen said, "we will see that doesn't happen."

Bob went straight to the airport and flew to Boston, where a high school friend had settled. He found himself, that night, walking on a lane in Cambridge, noticing the number 95 on a fence post. He opened the driveway gate. He stepped onto the driveway and turned onto a path beside a hedge.

When I first heard about this, and learned that the house whose gate he had opened was a historic home, once belonging to William James, the late-nineteenth-century philosopher and psychologist, I assumed that he was drawn onto the property because of its past. James had argued that the truths we consider to be absolute or objective are actually only pragmatic; we are certain about them simply because they serve us well. He proposed, too, that "religious genius," by which he meant both sensibilities that are deeply mystical and minds that exist at other edges, should be seen not as fundamentally different but as magnified versions of more common psyches, providing the chance to contemplate who we all are.

I thought that Bob was lured in, inspired to open the gate, by a plaque bearing James's name, by whatever knowledge he had of James's work, by the desire for some proximity to this thinker. But I was mistaken. A dim awareness of James's place in the history of psychology may have played some role, but it was the number 95 that enticed him, along with the majestic and mysterious look of the portico, the terrace, the gabled windows, the pair of stately, sentry-like chimneys. Ninety-five had been the address of our childhood apartment building in Brooklyn, a home before all had unraveled. And here was the number on the fence post, its metal 9 and 5 catching the minimal light of the lane and catching my brother's eye, calling out to him and calling his attention to the grand and lovely house. Did this home hold something for him? Some message or directive? Some solace? He felt that it might. He perceived significance in the coincidence of numbers. It was a habit or facility he had, the perception of significance. The geographical coincidence connecting him to Joan of Arc was the most dramatic example. Might James have doubted any bright line drawn to distinguish between Bob's feeling that these convergences were speaking to him in some way and the belief most of us entertain that the coincidences that crop up in our lives might contain some guidance or affirmation?

He walked the path toward the portico of the front entrance. He moved along a wall and peered in through windows. There was a room of wood paneling with floor-to-ceiling bookcases. It wasn't clear whether anyone was home, but it was obvious the house was lived in. He felt that he belonged here as well, because of the number and the warmth of the wood and the wide fireplace.

From the sidewalk, a woman asked what he was doing. He told her: just looking around. She threatened that if he didn't leave she would call the police. He didn't leave. A Cambridge patrol car arrived, and he told the police the same thing: just looking around. He agreed to go on his way.

From the moment he opened the driveway gate, and perhaps even from the moment he decided to walk down a lane in Cambridge at night, things might have gone much, much differently had he not been white. The odds of another outcome would have increased drastically with every choice he made: to turn off the driveway onto the path, to stare through the windows, to ignore the woman, to answer the police in a cavalier way that could have been interpreted as disrespectful or defiant. He could have been roughed up. He could have made the wrong gesture and, already a suspected criminal for casing the house, wound up killed.

But the police allowed him to leave, and he continued walking along Cambridge streets and around the Harvard campus, through what he thought of as intellectually hallowed territory, and saw, through a bank of windows that was partially belowground, students at work in a lab. One of the windows was open, and he climbed in.

At this point, he may have been completely beyond reason. The attraction of the number 95 had a logic to it, if not one most of us would have followed to the extent of stepping onto a stranger's property in the darkness, but ducking through a window and entering a Harvard lab may be beyond explanation. What he told me was "I was asserting my dominion over myself." As I write, I am hearing him the way others might. These are the words of someone beyond comprehension, or someone who can be comprehended only as crazy. But then, as his brother, with all the subjectivity that implies, I am hearing his words as the rationale of a young man whose self-dominion had been, in his mind, denied by our parents and by the hospital and by psychiatry and, in effect, by society. In passing through that Harvard window, he was refusing his status and scrambling into one of society's main citadels.

He was arrested. He was committed to a state psychiatric hospital for observation.

While he was committed, our mother searched for him in Boston, in Cambridge. He had let our parents know where he was headed. I don't know exactly where she searched. She is dead now, and our father's mind is addled by a type of dementia that sometimes accompanies Parkinson's disease. I imagine that she begged for information from his high school friend who lived in Boston and that he had none. I imagine that she begged at desks in police stations and was turned away because he was not a minor.

She might have pleaded at hospitals, both regular and psychiatric, and gotten nowhere, whether because they were not authorized or because she didn't cast her net as wide as the state facility outside Boston where he was sent. Were it one of my children, who are now in their twenties, I would walk the streets in my every waking hour, so I picture her in Boston Common, walking past the statue of George Washington on horseback, walking past the Frog Pond, staring into faces, afraid that she might see him and not recognize him, that he could be among the unhinged and homeless on the park benches, that already he had been physically transformed, and that unless she forced her eyes to linger she might overlook her own son.

And because her eyes lingered, the faces scraped and gouged their way into her mind. Faces obscured by heavy beards. Eyes searing and others unseeing. Old men with spotted scalps, whose features she scanned longer than anyone else's, with an initial, shape-shifting, paralyzing feeling that this was him, now, in front of her, followed by the vision that this was his future.

She scoured Boston with a primal need not so different from that of a man I knew, Ellis. His search was for his mother. Her moods had been wildly erratic. When he was a toddler, she was sentenced to prison. A relative had raised him, and when she was released, he pleaded to live with her again. He was nine; he wanted his mother. The family granted his wish. "Everything seemed to come in waves," he remembered the period after their reunion. "When times were good, they were so good. She was so happy. She taught me to make French toast. I can still smell the cinnamon. But other times there was no food, and she wouldn't get out of bed, wouldn't speak." There was a fire in the apartment. There was a violent boyfriend who worked as a bouncer. She, too, was violent. "She knotted a bunch of sheets together and tied them to a heating pole and told me, 'Next time, if it's a fire, if it's a fight, just put this chain out the window and climb down and get out.' Even in her craziness, she was trying to take care of me."

But she couldn't parent him, and he went back to relatives, and eventually he understood that she resisted her medication, that she alternately accepted and refused it. She became homeless. She became, when he saw her occasionally, gaunt, then emaciated. She was likely off her medication when she jumped from the Brooklyn Bridge, killing herself during his freshman year in college. The last communication he had from her was a birthday card written in crayon in a child's ragged letters: "I'm sorry that I failed at being your mother." Signing the card, she called herself his aunt, in acknowledgment that she had been replaced. And for years, after her suicide, he sought her out among the homeless and mentally ill outside Grand Central Station and at a McDonald's in Manhattan where he played chess with a man who ranted to himself and beat Ellis every time. Ellis bought them meals from the food trucks in front of Grand Central or fish cake sandwiches, the chess player's favorite at McDonald's, and tried to learn how they'd landed where they were and who they'd left behind. If he asked enough people enough questions, he felt, if he talked with them long enough, he would find her.

Our mother scoured Cambridge. She sat on benches in Harvard Square, surrounded by the pigeons that the homeless feed, waiting for him to appear.

"I THINK YOU'RE a very good candidate," the psychiatrist said. "The other thing to consider is ECT, but that's more invasive, with more side effects. This is really the right way to go."

So David sat back in the dentist-type chair while a technician measured his head. Psilocybin having failed, he was about to begin something new: forty-two sessions of transcranial magnetic stimulation, TMS, a descendant of electroconvulsive therapy, ECT. Back in the 1930s, ECT had been developed as another of psychiatry's means of rebooting the brains of the mentally disordered. Sufficient electricity was sent into the brain to cause bodily convulsions, sometimes dislocating jaws and breaking bones, but leaving some patients, for neurological reasons unknown, becalmed and grateful. Muscle relaxants, given before the procedure, curtailed the dislocations and fractures, but another side effect, the likelihood that the surge of current would knock out portions of memory, was harder to solve. Yet adjustments in the rhythms and length of the electrical pulses sent into the skull gradually lessened the extent of cognitive damage, and nowadays one hundred thousand Americans undergo ECT each year, mostly for persistent depression. Seizures are induced six to twelve times over a few weeks. One-third of patients feel their symptoms lift for at least several months. Maintenance shocks are often needed. Around one-third are left with the permanent loss of pockets of memory.

The electrical current generated by TMS is more gentle; the idea isn't to manufacture convulsions, which remains the key part of ECT, but to stimulate an area of the prefrontal cortex or, in David's case, two areas, one to address his depression and the other his anxiety. Each would receive separate pulses, ten per second in the depression zone and a slower pace in the location that was, in theory, a source of his fretting. About his insomnia, the psychiatrist at the TMS clinic said that the two pulses had good odds of treating this, too.

The technician measured David's head with calipers. When the psychiatrist named the prefrontal areas that he would target, David lumped these regions together and relabeled them the hooby wa hooby. He wanted less information, not more, and not only because he was in a state of cognitive disrepair that left him cowering from his part in potentially historic legal work but also because of what he'd read about TMS. Some writers cited data on its uselessness and likened it to snake oil; others pointed to statistics of great success. He had given up on evidence and, though he was not religious, heard himself saying or thinking sentences that began with "I'm just praying that . . ." or *Just pray that after this . . .*

He was relieved that his first visit wasn't taken up solely by the measuring. The technician lowered the conductor and positioned it against the side and front of his head. She asked what he wanted to watch on the wall-mounted TV. He chose *The Office*. She turned on the TMS machine and left the room. The sensation was of a woodpecker hard at work inside his skull. For forty-five minutes, he watched Steve Carrell and the other employees of the paper company clash in their deadpan comedy while his brain was pecked.

That night he slept six hours. This was stunning. At his session the next morning, he told the technician about his breakthrough. She said it was to be expected. "We just exhausted your brain. It won't last. The lasting effect will take a while." She situated the concave surface of the conductor against his head and switched on *The Office*. The second night brought regression, and the third and fourth sessions brought no change, and during the fifth, despite the woodpecker and the comedy, he nodded off, as he frequently did during the day from his mix of fatigue and despondency. The technician came in to tell him to stay awake, that otherwise the treatment wouldn't be effective. The psychiatrist assured him that he might not notice the benefits for three to four weeks. Toward the end of the second week, he heard himself laugh aloud at something on the screen and asked himself how long it had been since that kind of laugh had burst from his throat.

He told the psychiatrist and the technician that he'd like to ramp up the stimulation to full intensity as fast as possible. For caution's sake, they preferred to dial up by increments. David insisted, and the psychiatrist relented. After that, he felt new energy and engagement, driving across the city to pick up a stack of postcards, preaddressed to voters around the country, then driving home and adding a personalized note to each one, reminding people about the importance of both the national election and their local races and urging them to get to the polls. He made a selfie video to commemorate his progress and to play for himself if he slid backward.

He did slide. The better days seemed a sadistic illusion. He asked the technician how often patients got better, to tell him again. The psychiatrist asked how he felt during the sessions, and, when David said, "Nothing worse than uncomfortable," the psychiatrist asked if he'd like to go beyond the standard range. "The greater the intensity," the psychiatrist said, "the deeper it penetrates." The woodpecker's beak struck hard. David left the clinic with his head throbbing.

He told no one about this aftereffect; he dreaded being put back to a previous level. Four weeks in, he asked the psychiatrist whether, if the prescribed forty-two sessions weren't enough, he could go on. The psychiatrist started the process of getting approval from David's insurance company. David drove home from the clinic each day, accustomed to the throbbing but never to his parrot-like voice demanding his dismal updates and eliciting the defense that sounded more and more feeble to David himself, *This is not me.*

And yet two hopeful things occurred during this time. One was that Biden's lead widened. David knew better than to count on the margin holding up, but he let himself imagine that Trump might lose so convincingly that he wouldn't put up a fight, that there would be no need for legal activism. The second thing stemmed from Gillian's planning a sleepover for an upcoming Saturday night at the house of a basketball teammate in her pandemic pod. Amanda said that she and David should drive to the mountains, hike before the season's first serious snowfall, and spend the night at a hotel. She'd already researched hotels with COVID-vigilant policies. They were going to have a romantic weekend.

"THERE WAS AN iron staircase on the side of the building," my brother recalled recently. "That was the entrance to the homeless shelter. It was called the Bristol Lodge. That was where I lived."

After his arrest, he had been involuntarily committed for observation to determine whether he was competent to stand trial for breaking and entering. Except for the locked doors, the state facility didn't have much in common with the university hospital psych ward where he'd been in Seattle. Built in the 1920s to house a thousand patients; built on a monumental scale with Greek columns towering before the entrance; surrounded by gardens and a graveyard to bury those never claimed by their families; equipped with special rooms and surgical theaters where the procedures of each successive era were performed, the facility was, by the time of Bob's stay, decrepit and three years away from being shut down. He walked dingy corridors with men condemned to the shuffle of Thorazine and Haldol. But his psychiatrist there was kind, and though he counseled Bob to go back on lithium, he made no legal move, demanded no hearing, to impose the drug.

The psychiatrist deemed Bob competent. After two and a half weeks at the facility, he was cuffed, loaded into the back of a police van, and taken to a holding pen below a courthouse. The pen was mercifully underpopulated that morning, though one man was lying on a bench and masturbating.

A while later, an officer led Bob upstairs into the courtroom. There, at the defense table, a public defender let him know that the young assistant district attorney was eager to put him away for breaking and entering with intent to rob, a felony. But the prosecutor was lacking a witness; the Harvard student he'd counted on was apparently far away on an exotic vacation; the prosecutor was frantic and asked for an adjournment. The judge made his ruling. He turned to one of the officers and said, "Give him his belt back."

Before they parted, the public defender told Bob a story about his own father, who'd been manic-depressive and who'd killed himself. He advised my brother to think again about medication and asked what he planned to do next, where he planned to go. Bob said he wasn't sure. He confessed that he had no money, no belongings, nothing, that his wallet and bag had disappeared somewhere between landing in Boston and the psych hospital. The lawyer gave him five dollars and the address of the Bristol Lodge men's shelter. Bob reclaimed his belt from the officer. He slipped it into the loops on his pants for the first time since he'd relinquished it at intake. He stepped out through the courthouse doors and walked to a bus stop and rode to the address the lawyer had given him.

"It was in an old firehouse. Everyone had to be out during the day, so it was getting dark when I checked in. Before you got to the front desk, you passed by a common room with guys sitting around a TV, and big plastic garbage bags of donated clothes for guys to rifle through. But the person behind the desk was, 'Hey, welcome, we'll get you a bed, sure,' all upbeat and matter-of-fact. He showed me around. Bathroom and showers. Two rooms of beds, fifteen to a room. He talked about how sobriety was a house rule. He was, 'Make yourself at home.'

"I didn't feel any trepidation. Which might have been delusional of me. But the sheets were washed, the place was clean, I was thinking, *One step at a time, this is all going to be fine, this is all going to work out.* I was feeling more freedom than fear. I rummaged through the bags of clothes and found a pair of tan corduroy pants and a white dress shirt, and in the morning I went to the local employment office. I said I could cook. Which was based on a job I had at a juice bar in Seattle. And in a couple of days, I was cooking at a nursing home. I had two shifts a week.

"On the days I didn't work, I lined up for dinner outside a church across from the shelter. That was something. The guys who slept next to me, who spent their days in the park, they wandered over, and we all waited on the street to get in, waited to be fed. We filed into the church basement and took plates of spaghetti and chunks of bread. The basement was too crowded for anyone to sit alone, but I didn't talk to anyone while I ate. Most of the guys didn't. For those men, there was a lot of solitude. Afterward I sometimes watched TV. Tiananmen Square was happening, and I talked with people a little about that. But mostly I read a biography of Dwight Eisenhower that I found in the nursing home library. I loved that book. The idea that Eisenhower knew D-Day would probably be a calamity but that he went ahead with it because he had no other choice. There was a quote of his that I've never forgotten. This probably isn't exactly it, but for me it became sort of like a credo. 'We did what we could with what we had when we had it.' I knew how anyone would see me. I was a guy lining up on the street for my dinner. I was a guy taking my wrinkly white button-down shirt into the shower with me to wash it at night. I was one more homeless person hard on his luck.

"But I didn't feel hard on my luck. I felt like I was escaping from this crushing fate and that whatever happened was going to be better than what I was leaving behind. I was escaping the idea that there was something wrong with me, that I needed to take drugs, that even with the drugs I couldn't trust my own judgment, that I was broken, that *you're fucked, you're fucked, you're fucked, you're fucked.* I thought—not for the first time, but this time it was a clear calculation—*I'd rather be dead than be a broken person.* And even though I had just been arrested, and even though I had just spent weeks on a psych ward for the second time, and even though I was going to be living indefinitely in a homeless shelter, this was better than the other narrative. That was a life I just did not want to live."

TEN

Two parables and a line from a song cast light down a path that diverged from conventional psychiatry. One parable came from a man named Chacku Mathai. The song lyric was spoken by a neuroscientist, a colleague of Nestler's at Mount Sinai and research director with the U.S. Department of Veterans Affairs. The other parable served as a statement of principle for a psychiatrist who had turned away from a traditional and esteemed career:

> Once, there was a prince who went crazy. He believed he was a turkey. He took off his clothes and crawled under a table and refused to eat anything except to peck at bones and bits of bread, which were left for him on the floor. The king and queen were distraught. They called for the royal physicians, but one after another failed to carry out a cure. The prince remained a turkey, naked and pecking.
>
> Then a sage arrived to offer his services. He undressed and sat beneath the table, nibbling.

"Who are you?" asked the prince. "What are you doing here?"

"And you?" asked the sage. "What are *you* doing here?"

"I am a turkey," said the prince.

"I, too, am a turkey," said the sage.

They spent time together like this, getting to know each other, until one day, the sage asked the king's servants to bring him some shirts. "Is there any reason why a turkey cannot wear a shirt?" the sage asked the prince. Each put on a garment. The sage signaled the servants for pairs of pants and asked, "Is there any reason a turkey cannot wear pants?" The same transpired with socks and shoes, and soon they both were completely dressed.

Next, the sage requested regular food and, when this was brought, asked of the prince, "Is there any reason why a turkey cannot eat good food? I think one can eat what one wishes and still be a turkey."

They feasted together, and sometime later, the sage inquired, "Is there any reason why a turkey must sit under the table? Surely a turkey may sit at the table, with his place nicely set."

And in this way, the prince was fully cured.

Pesach Lichtenberg, who found professional wisdom in this rabbinic tale, had grown up and gone to medical school in Brooklyn and the Bronx, then moved to Israel and finished his training as a psychiatrist in the early '90s. He'd been swept up in the psychopharmacology of the time, and once, as he made rounds with a senior colleague, he had listened to a patient speaking, he told me, "about God and demons and the messiah and so forth. I was fascinated. I've always had the problem of being intrigued. I was getting lost in his descriptions. But as we walked away from this person, the senior psychiatrist said, 'That's not him. That's

his dopamine talking.' It struck me as such a wonderful insight."
Lichtenberg laughed almost bitterly behind his cropped goatee.
"Today I'm ashamed that I could think this way."

Lichtenberg was wry and slender; it wasn't hard to imagine
him folding himself up beneath a table with a patient. But for the
majority of his career, that's not at all what he did. He presided
over a psychiatric ward at a Jerusalem hospital. We visited his
former workplace not long ago. The hospital at first approved and
then rejected his request to show me around the ward, but we did
descend within the building to a semi-underground level, where
we walked a series of long, utterly undecorated corridors that led
to the locked doors. While he chatted with the current director, I
looked through a window at the bunker-like area where patients
received their outdoor time. Two men occupied a pair of concrete
benches. One sat hunched and torpid, the other lay on the concrete in a fetal position.

For over twenty-five years, Lichtenberg kept men like these
sodden with medication. "Half the dose was to calm the patient,"
he said, "and the rest was to assuage the anxiety of the staff." But
even so, staff felt his dosing was insufficient. "I discovered later
that the legendary head nurse of the ward, a survivor of Ausch-
witz, thought I was being too gentle and would take matters into
his own hands. He would supplement. The head nurse of the
whole hospital finally led me to understand: he's been giving injections behind your back, because you don't know how to treat
your own patients."

As Lichtenberg grew thoroughly disillusioned, in the mid-
2000s, a new patient was brought to the ward in a state of acute
psychosis. Avremi—Abraham in English—was in his twenties, a
devout student of kabala, an ancient form of Jewish mysticism.
Weeks earlier, Avremi had purchased the thirty-seven fruits that
symbolize the kabala's four worlds, culminating in the seven
pure fruits that are most divine. Building a meal around these,

he celebrated a sacred holiday with kabalistic friends, and that night and over the next days he began to feel, he recalled when I met with him, "that my legs were a little bit less on the floor. Everything was in the hands of God except belief. And maybe even belief. Only one rope held me to the ground. And then that rope was cut. Everything was God; it was like I entered a quiet light."

His new wife, Yisca, the daughter of Yemeni immigrants to Israel, was not pleased. She and her family were at least as devout as Avremi—but this was scary. "Things started to go more and more," he said. He had meetings with a prophet from the Bible and with Nebuchadnezzar, the Babylonian king who had destroyed Solomon's temple in the sixth century B.C.E. He consulted a revered living rabbi, asking whether, as he suspected, he had been chosen for a pre-messianic purpose. "The rabbi closed his eyes; he was thinking. And then he told me that it was not so. I was very disappointed; I had felt I was on top of the world; but I accepted what he said at that moment." Avremi contemplated, though, what another rabbi had taught him, "that if one person, just one person, *wants* the messiah to come, he will come. So I thought, I may not want more than anyone else on the planet, but I want to want in this way, and I will pray from this position—like from a sanctuary of the heart—the position of wanting to want."

He told his wife he was going to Jerusalem's Old City, to the *kotel*, the two-thousand-year-old wall that touches the remnants of the second Jewish temple, built by Solomon's descendants. Because of this proximity, the wall is the holiest place of prayer in all of Judaism. Avremi went, dressed in white. He also carried a flowing white tunic. This was for the messiah to wear, when he appeared.

Avremi prayed at the wall of rough limestone blocks. It was a bright, warm day, and the plaza in front of the *kotel* was filled with all manner of ultra-orthodox worshippers in their long black or embroidered coats, the socks that distinguish their sects, their fedoras and fur hats, alongside non-orthodox visitors who never-theless were moved to press their lips to the stone and to write words of prayer on scraps of paper and slide the folded scraps into the wall's cracks and crevices. Avremi prayed and prayed. He joined a group of orthodox children and prayed with them, because the prayers of children would assist in his mission. The messiah did not materialize. He thought about the determination of Joshua, the biblical warrior, as he led the Israelites in their attempt to con-quer the promised land, and he lowered himself to the stone floor at the foot of the wall. He would not give up. With his white prayer shawl draped over his head, he rocked back and forth in time with his incantations, a white mound of extraordinary devotion even by the standards of the ardent believers surrounding him. He sat on and on, prayed on and on, hour after hour after hour, and then—vaguely he knew this was happening—he was being lifted into an ambulance.

He was put in an isolation room, diagnosed as psychotic and perhaps afflicted with catatonia, a disorder marked by stupor and immobility that can accompany schizophrenia. He was about to be treated with an antipsychotic and a benzodiazepine. By that point, it was morning, and Lichtenberg arrived at work. He had Avremi brought into a meeting of staff and psychiatrists-in-training. "I felt like a mouse," Avremi said. "Or a rabbit. Or a rat. Like, *Let's see this psychotic person, let's see his sickness.* But at the same time, I felt that one of them had a good voice, that he had empa-thy, that he honored me. I felt, because of this, my eyes start to get a little bit wet. Before that, I was in so much stress and in eupho-ria, both, I wasn't feeling anything, I was numb, I was thinking I will give my life if I must to bring the messiah. But the empathy in his voice made me start to feel my feelings."

Lichtenberg, a religious skeptic who had been raised in one of Brooklyn's orthodox communities, arranged for a family rabbi of Avremi's to come to the ward. The rabbi conducted a ritual releasing Avremi from his vow to bring the messiah, and Lichtenberg took part, a kind of enactment of the turkey prince parable. He sent Avremi home without pills, without a prescription, without even recommending that he consider medication.

During this period, Lichtenberg was beginning to have his own visions. He'd read about the Soteria Project, a pair of houses in California's Bay Area that offered psychotic patients, in the 1970s, an alternative to hospitalization. Founded and run by Loren Mosher, a psychiatrist who led schizophrenia research at the NIMH but became appalled by the widespread use of antipsychotics, the Soteria houses dispensed with hierarchies of knowledge, so that professionals like Mosher himself were seen as having no more expertise—and perhaps less—than the people they were treating. The homes relied on minimal medication and enshrined two words, "being with," as their treatment philosophy, meaning that the staff, who lived in the houses, were expected to converse and commune and coexist with, rather than govern, the residents.

Over twelve years, Mosher's project served two hundred or so men and women, who would stay for six months, before the experiment collapsed for lack of funding, and Lichtenberg, reading Mosher's book about Soteria, was enthralled. Mosher wrote that seventy-five percent of his residents either never or only briefly received antipsychotics and that, in comparison to a cohort of non-Soteria patients who served as his control group, "all Soteria residents did significantly better two years after entry to the program on an eight-variable global outcome measure." Reading, Lichtenberg felt that he himself was being delivered.

He wrote proposals and peddled the concept of Soteria in Israel to the ministry of health and the country's public system of health insurance carriers. It wasn't only Mosher's story that gave him confidence. The World Health Organization, the WHO, had done a series of large studies, between the 1960s and '90s, comparing long-term results for people diagnosed with schizophrenia in nations classified as "developing," such as Nigeria and India, where antipsychotics were scarce and less frequently used, to outcomes in the United States and Britain and other "developed" countries. Patients in the less-medicated nations came out well ahead. Their odds of remission were daunting—around forty percent stayed largely free of symptoms over a five-year period—but were twice as good as in the wealthier countries. And in terms of social functioning, outcomes were much better in the less-developed nations. Critics pointed out that data isn't easy to standardize across cultures, but it was difficult to dismiss the fact that decades of research had produced consistent results.

Research done solely in the West, rigorously comparing medicated to nonmedicated outcomes, was hard to come by, because the profession deemed it unethical to withhold antipsychotics. Yet a Dutch psychiatrist had tracked over a hundred patients for seven years, all of them medicated during their initial episode of psychosis, after which some were randomly assigned to eliminate or markedly reduce their drugs. This cohort fared better than patients who were kept on normal maintenance regimens. Several studies of humans and primates suggested that protracted use of the medications caused a degree of atrophy in the brain. The data was debated but haunting. Nancy Andreasen, whose *The Broken Brain* had hailed antipsychotics in 1984, summarized in 2008 new research she had done on the medications: "The more drugs you've been given, the more brain tissue you lose."

As he envisioned his own Soteria, Lichtenberg could have worried about the reputed connection between psychosis and violence, but he drew reassurance from data spanning eleven countries, showing that the link to violent criminality was almost entirely due to factors like substance abuse and poverty, and only marginally to the disorder itself. Yet some studies were less sanguine. Taking all this in, Lichtenberg's mind ran to the stories we tell ourselves, the realities we construct. Most of us hesitate or refuse, on moral grounds, to brand subgroups of people with elevated rates of violence. We condemn such categorical thinking, because of the discrimination it produces, and because it precludes seeing people as individuals. But with the psychotic, we're quick to tell ourselves a violent tale. The media helps us along. It reports any speculative assessment that a perpetrator is mentally ill. Even if the media's intention is humane, our fears are perpetuated. "If the reality was anywhere close to the perception," Lichtenberg said, "the data would be much more clear."

Lichtenberg drew, as well, from his own work at the hospital. When patients became menacing, he told them how frightened he was. "Mission accomplished for them," he told me.

But year after year, his campaign to win over the health ministry and the insurance carriers got him, at best, words of encouragement—no licensing, no financing. His Soteria plans were wafting. Then he reconnected with Avremi, who, eight years after his effort to woo the messiah, had earned his undergraduate degree. Now he needed some supervised experience to gain a master's in psychology. He thought he should try going back to the hospital where he'd been committed and back to the psychiatrist with the empathic voice. Lichtenberg gave him the chance, and, Avremi said, "I started to make trouble. I joked with the patients, I danced with them, I unlocked the door to the isolation room."

To Lichtenberg, though, it was plain not only that he'd made the right choice in releasing Avremi years earlier, but that he was right to welcome him back. Above his Hasidic beard and behind his wire-rimmed spectacles, sly joy and soothing gravity blended in Avremi's brown eyes. His words were both earnest and self-deprecating, self-amused. He bonded with chronic patients. He listened to Lichtenberg's Soteria dream and told him he must not delay. And around the time Avremi started prodding him, the media latched on to the story of a woman who'd been strapped to a psych ward bed in Jerusalem for three weeks straight. A private foundation took an interest in Lichtenberg's notions; the health ministry gave what seemed to be concrete encouragement— though this evaporated; and in 2016, he decided to proceed, unlicensed, with the foundation's start-up backing. He rented a house next to a matzoh factory that baked—and warmed the adjoining wall—from Chanukah to Passover. He appointed Avremi his director, and Israel's first Soteria house opened its doors.

Within moments of stepping through those doors, a few years later, I fell into conversation with a roving American, a rapper named Shmuel, with a light brown ponytail, who'd landed in Jerusalem, then landed here. "I invented the term 'craepride,'" he said. "Yeah, my brain is different, but ladies love brains that think differently." He stood from the threadbare sofa we shared in the vestibule, disappeared, and returned with a bound thesis that had recently been rejected by an Israeli college he'd been attending. "It's called 'Jeudaemonia,' which is derived from 'eudaimonia,' which is Aristotelian. They said I didn't cite enough academic sources, but look—suck my dick!—I quote Plato, Maimonides, the Zohar, I quote Eminem, Macklemore, and myself." He wore a New York Islanders tee shirt, emblazoned with a hockey stick, and slid without so much as a split second's pause into his own high-speed rhymes:

> *Mental health plunging*
> *Neurons defective*
> *Hero of the generation*
> *With neurodegeneration*
> *Time on earth limited*
> *Head sick, death imminent*
> *Extra-existential crisis*
> *Shit's poetic, isn't it?*

An hour or two went by, and I was still sitting with him; the vestibule was as far as I'd progressed into the house. This was the way Soteria was supposed to work on people—residents, staff, some of them with psychiatric histories, former residents who came to visit, family, friends surrendered to the space and sought a communion of minds. Shmuel had gotten here via a punch through a window and a stint on a Jerusalem psych ward. "At the hospital, they put you under fear. *If you don't take these pills, we're going to hold you down and give you a shot and turn you into a zombie.* That was the word they used; 'zombie' is the same in Hebrew, with a distinct inflection in the pronunciation. At Soteria, it's okay that your brain is different; it's, let's learn how to live in the world with this brain. It's not quite celebrating the craziness, but it's celebrating the neurodiversity."

Up a half flight of stairs was a courtyard with benches and a living area with a vaulted ceiling, a Ping-Pong table, a battered piano, stray soccer balls, two guitars, a darbuka—a Middle Eastern conga drum—soft, secondhand furniture, and an open kitchen where challah was being braided. Above, an internal stone balcony floated over the scene, graceful but bearing a hill of house laundry.

In the courtyard, three residents and two *melavim*—companions—sat around a backgammon board. The house tended to hold, at any given time, ten residents, one or two part-time psychologists or social workers, a director, who had been Avremi and was now a therapist with an equally unusual journey to a career in psychology and to this place, and a few *melavim*. The *melavim* were paid interns except that in Soteria they were exalted as an essential part, possibly the most essential part, of the treatment, because they were responsible for establishing the atmosphere of coexistence, of genuine curiosity, of asking, of being with, of desiring to understand and engage with each resident's reality. They played backgammon casually, idly, as if they were home for a holiday and passing time in their family's living room, while they chatted with the three residents, one of whom mentioned Descartes's proof of existence, "I think, therefore I am," as a source of his troubles. He hadn't studied philosophy in college; he hadn't been to college. But one night, as he worked as a security guard and passed the hours in his booth playing video games, he somehow wandered to a website that included Descartes's dictum. "It is stuck in my head like glue," the young man said, his lips forming a stiff, painful smile and his eyes in anguish below his bangs. "When I think, 'I think, therefore I am,' it means that I'm here but not anything else."

Before that night, he'd had issues; he'd been prone to obsessional thoughts. But since, because of Descartes, everything outside him was unreal. The *melavim*, the other residents, the backgammon board, none of it existed. He had also watched *The Matrix*. The sci-fi film had worsened his isolation. He knew his mind was awry but couldn't set it right. He could touch the benches, the board, us; it didn't help. "Like glue," he repeated.

In the living room, Shmuel picked up a guitar and switched from rap to oldies rock, strumming the Eagles' "Take It Easy" and singing loudly in his scratchy voice. Bean stew, hummus, spaghetti Bolognese, and the challah, all made by residents and *melavim*, were set out on a dining table. Some residents gathered for lunch, two were out for a walk through the city with one of the *melavim*, some took their plates to solitude, one or two remained in their rooms. Lichtenberg's ambition was to operate with almost no rules, certainly without guidelines imposed by him or his professional staff. He liked to say that his only requirement was that fresh hummus should always be available and that it should be made according to his own well-honed recipe. As we ate, a past resident told me about trying to teach two fellow residents to paint while he stood naked in the living area. "One of the *melavim* asked, 'What are you doing?' Afterward, I looked in the mirror, and I said, 'You are psycho,' but the *melavim* allowed me to do it." He had conducted his class in the nude.

Shmuel's craepride, the Cartesian's awareness that the unreality of all around him was both real and not real, the naked painter telling his reflection, "You are psycho"—their knowledge of their own divergence seemed to call into question a key concept in the psychiatry of psychosis: anosognosia. It meant the lack of awareness of one's own condition. Yet, from Caroline to the painting teacher, my talks with those diagnosed as psychotic were filled with a dual cognizance; they knew that my reality wasn't theirs. Perhaps what the profession heard as a lack of awareness was actually a rejection of clear lines, of psychiatric delineations that were like life sentences. As best I could tell, the psychotic, much more often than not, held two truths in mind.

One evening, I sat in the alley outside the house with a man who was under surveillance. The Old City was a short walk away in the darkness. I had just spent a few hours along its cobbled lanes, swept along by its crowds, listening to the bells of its ancient churches and the enchanting drone of its muezzins and the murmur of its Jewish faithful bent over their siddurs. I had inhaled the living scent of the icons in the Ethiopian monastery, and I had walked amid the tombs of Mount Zion, where, one day, the bodies will rise. Belief was as prevalent as oxygen in the air, yet even the most devout also navigated the world I inhabited, subscribing to multiple truths.

"The surveillance company knows my every inner private thought," the man said. He was among the most troubled I talked with. "There is no boundary around my mind. They want me to eat from the apple of my exposure. If I knew the company was good, I would be okay, but I don't know. Groups of civilians are fighting and fighting, and water is flooding the bridge, and the company is asking me to take part in the management of the warfare. You can sit here and speak with me and then go home and think, *I had this day*. But I can't."

Lichtenberg had opened a women's Soteria across town. At the men's house, the staff and *melavim* were male and female; here all were women. Yet a new resident was sleeping in the walled garden in a tent, because she felt safer that way. Ori, the social worker in charge, said that safety—and the fragility of it—preyed on the minds of many. "I am thinking about another of the women; she also communicated that she was terribly frightened. 'Because what if someone will come in the room and rape me?' I told her that we always are very careful, and she accepted this, but, after a few seconds: 'What if someone *wants* to rape me? You can prevent them from doing it, but *wanting*, what about that?' So we talked about this problem. And then she asked, 'What if someone wants to rape me, and they don't even know it—that affects me, too, even if it's subconscious.' This whole conversation was with our eyes deeply in meeting, and she asked, 'What if I want to rape you? What will you do then?' She wasn't flirting, and she wasn't talking nonsense; she was posing these core questions about human relations: What if you want something from me that I don't want? What if I want something from you that you don't want? She was saying that there is so much happening beneath the surface of human existence, so much that is barely controlled, and that it is tearing away at her inside. I wouldn't want to be in her state, but psychosis can allow these things to be expressed. I had to say, 'Wow. These are good questions,' and that let her feel, *Okay, it's safe here, I can go through this here.*

"For the *melavim* and staff, the difficulty is not being unable to see yourself in the residents. It is that seeing yourself is too easy. For some, this is scary. You put up your defenses. You push the knowledge away. You have to learn that it is okay to be scared. And then you can agree, *This is similar to me.*"

This was the world of Soteria, and in Israel, Soteria was expanding. There was a third house in the north of the country, a fourth and fifth in the works, and a handful of new, closely related homes that didn't bear the Soteria name but shared its approach. The latest to open was the first to cater to Arab Israelis. The ministry of health had licensed Lichtenberg's homes, and two of the country's four main insurance carriers had signed on and would pay for a stay in Soteria as an alternative to hospitalization. So would the ministry of defense's own insurance system. The Israeli media had run laudatory stories. Two public hospitals were talking with Lichtenberg about opening Soteria-like units. Lichtenberg's hope was to have his houses multiply into a movement that was revolutionary in scale. His dream was to have, one day, over a thousand residents—roughly one-third of the number of psychotic patients presently on the nation's locked wards—being treated at Soteria homes and their offshoots. As a step in reaching this goal, he enlisted Avremi, who was working toward his psychology doctorate, to help him in drawing up a list of techniques, so that he could reply coherently to the question that often came at him: "But what exactly do you do?"

It was a nearly impossible question to answer, because there were no prescribed methods. Group discussions and sessions of drama therapy did take place, but they were secondary to the effort at ceaseless empathy. The list he and Avremi composed contained items such as "blur the boundaries between 'normal' and not" and "judge not and shrink not back." The *melavim* and staff sometimes asked for specific rules and guidelines, sometimes begged for them, but Lichtenberg believed that these would inevitably become barriers, impediments to relationships, to listening, to letting go of resistance, to insight, to ingenuity, to discovering and offering whatever was needed. "I tried to imagine what it was like to be in his world," one therapist said, remembering a resident. "I could do this for maybe a minute before it became overwhelming, but for him, it was one minute and another and another and another, and I could only try to be compassionate and soft, to help him find some quiet in his agitated mind. I have three little kids, and at night I sang to him what I sang to them, with my hand on his shoulder, while he lay in bed, the same lullabies."

One resident laughed randomly and, when he wasn't laughing, examined the living room walls, touching them like reading braille, as if mesmerized by something within or behind. "He was so disconnected," a *melaveh* said. "We thought he was lost, permanently. Anything you would ask him, he would say, 'What? What?' He would let no one inside. But I had found out he played guitar, and one day I played, and I asked if he would pick up the other guitar, and I played four chords in a loop, and he improvised, in tempo, on the scales. He was talented. I wanted more, more of this, more of his jamming. He was sitting on the sofa and I was across from him and another resident came with the darbuka drum, and the first resident, he said something about feeling trapped, and I riffed on that, singing something with that, and he got really into it, bending the strings, and we alternated—I sang and he soloed, I sang and he soloed.

"After that, he couldn't ignore us. He couldn't shut us out. He couldn't stay at the walls. When we asked him how he was doing, he would say, 'I'm okay, how are you feeling?' He started to share in the cooking. He shared that he does hear voices. Once he had a safe place in consensus reality, he could start to share what he heard."

One of the offices in the men's house had an internal window that overlooked a cistern from centuries ago. It was rubble now. Sometimes a resident asked one of the *melavim* or staff to climb down there with him, to talk there, in the dirt and darkness, or to hurl rocks at the remains of an underground wall, as if to attack the imprisonment he felt within his mind. Whoever was asked ducked through the window and descended, following where the resident wished to go.

So much about Soteria seemed magical. When I asked Lichtenberg about eruptions of violence, he said that chairs had been thrown and broken, plates smashed, a table destroyed. And though Soteria had a liberal policy about leaving for hours-long outings with *melavim* or friends—something I'd seen when I went out to play basketball with four residents and one *melavah* in a local park—the front doors received regular kicks and batterings. But threats against other residents, *melavim*, or staff were rare and, with the exception of one broken nose, had ended with hardly more than a scratch. "If someone becomes intimidating," Lichtenberg said, "I'll sometimes put my hands behind my back, look him in the eyes, and tell him, 'If you want to attack me, it's going to be so easy for you.'" Avremi spoke of a resident who, on his first day, had stepped abruptly away from a group meeting. He punctured the head of a darbuka, bashed and splintered the drum against the floor, and danced furiously. In response, Avremi joined him, dancing wildly, fiercely. They danced together in the middle of the group circle. "Everybody made a drumbeat with their feet, stomping, and we fought, in a choreographed way, a dance-fight. He grabbed me; he put me on the floor; but I wasn't hurt; and later we talked, and we went down into the cistern. He was asking what was happening to his mind. He was crying. I slept near him that night, and when he woke with nightmares I sang him songs, like a mother with a baby, and gave him tea."

Yet in one way, Soteria was falling short of magic. At the time I was there, Lichtenberg hadn't compiled any data on his prescribing of medication at Soteria; a year later, Avremi put together preliminary numbers as part of his doctoral research. The numbers didn't come as a total surprise to Lichtenberg. He knew that over Soteria's five years, there had been a shift. Initially, he had sustained a fair degree of purity, adhering to his stance against antipsychotics, and some of this spirit persisted. People like the rapper and the painter and the man under surveillance were free of antipsychotics when I visited. But Avremi's data showed that gradually Lichtenberg had given way.

The current pattern was this: There were a minority of residents who arrived unmedicated, frequently because they were in the midst of their first psychotic episodes, and with most of these, he didn't resort to coaxing them toward medication. But as for the majority, who came to Soteria already on antipsychotics—and who almost invariably had been hospitalized in the past, probably given bludgeoning doses and released on maintenance regimens—he no longer tried to wean them off their pills. They tended to reach Soteria's doors in agonized straits or because their families feared dire emergencies; no one was seeking treatment when their minds were relatively quiet. And since the insurance carriers restricted their stays to five to twelve weeks, Lichtenberg didn't have the time he felt he needed—five or six months—to try incrementally reducing or eliminating the drugs that had already caused mechanical and structural alterations within their brains.

This was one explanation for Soteria's unspoken shift in policy. But another possibility was that with the approval of the ministry and the funding of the carriers, Lichtenberg had lost some of his ardor. It was harder to be a revolutionary when he had won such conventional endorsement. He worried that he had grown more timid about venturing away from the mainstream. But he also worried that antipsychotics, awful as they were—one recent resident I met with suffered telltale hand-flapping, and another felt the torturous jitteriness of akathisia—might nevertheless prove unavoidable in many cases. "I had a rather inchoate fantasy," Lichtenberg said, recalling Soteria's first months, "that there would be so much support, so much love, who would need medication?"

For the future, he hoped that his houses would have more freedom, more funding and trust from the carriers, more time to truly test the influence of empathy, of being with. Still, for the moment, he was engaged in a rearrangement of perspective, a reordering that relegated chemistry and medication to secondary status, a revolution of scrambling down into a cistern, of singing lullabies and eliciting solos, of dancing as if violently with a resident, of listening to a man poisoned by Descartes, a man for whom you were not real, and instead of thinking, *But I am*, attempting to imagine yourself into his utter isolation, and joining him there.

CHACKU STRADDLED WORLDS and worldviews: East and West, radical activism and minor reform, an oceanic consciousness and a job with Goff's wife, a Columbia University psychiatry professor, Lisa Dixon. His parable went like this:

> Once, long ago, a traveler to a new land came across a peacock. He had never seen this type of bird before, and he took it to be a genetic freak. He felt pity for the poor creature. He believed it could not possibly survive in such a strange and deviant form. So he began, carefully, charitably, to correct nature's error. He cut off the long, colorful feathers. He cut back the beak. He dyed the bird a speckled, muted black, making it resemble a species that was familiar to him. "There now," he said to the bird, with pride in his work, "you look almost like a regular guinea fowl."

Chacku's parents were immigrants from riverside villages at the southwestern tip of India, and in the early 1970s they landed with their two young children in Rochester, New York. The city's black population had grown by five times since the '50s; rioting had broken out in the '60s after the aggressive arrest of a black man at a street dance, rioting that was put down by police dogs and the National Guard. When Chacku's family settled there, it was a city of racial hostility and fear that jobs at Rochester's biggest employer, Kodak, would be lost to outsiders. Chacku recalled his father, a medical technician, gripping his hand tightly on the street when it was time to speed up their pace. He sensed spite and danger surrounding his parents. When he volunteered, in kindergarten, that water was called *vellam* in Malayalam, the language he spoke at home, his teacher scolded him, informing him that he was in America and telling him, "Speak English!" There were beatings in a first-grade coat room and, later, punches to the stomach followed by kicks to the head after he was knocked to the ground at recess. Retaliating, he pounded a boy's head into concrete and one day after school swung an iron bar at kids who had scorned him, striking no one but sending them into flight. He was clobbered with a baseball bat above his eye and stitched up at a hospital, and he was attacked in the locker room showers at school, battered to the floor, and dragged naked across the tile.

Early on, his school recommended to his parents that they rename him to help him get along and make things easier on everyone. He became Terry and was soon called Terry the Fairy. In a play on his last name and because he seldom spoke, he was also known as Too-shy Mathai—or Zizz, because when he did speak, he mumbled and stuttered. Due to his fighting, he was placed in special ed. In sixth grade, he pulled a knife on a student who threatened him. To get to his classes, he circled outside the school building, so he could avoid the taunts that came at him constantly in the hallway, though something disorienting was happening: He couldn't always tell whether the taunts were real.

Something was happening at home as well. It involved his mother's curry, his favorite dinner. With her hair in a single braid that fell to her waist, and wearing an ankle-length white gown decorated with red flowers, she grated fresh coconuts in a kitchen of cramped counters. "She would ram and rub the chopped coconuts over a big spoon with a serrated edge," he said. "The spoon was part of a thick block of wood that sat on the floor. She'd brought it from India. And she chopped the chicken with a giant cleaver. That's how you knew it was going to be chicken, which I preferred to beef—you'd hear the hacking, the chicken bones breaking. There was no rhyme or reason; it wasn't cut a breast and cut a thigh; it was chop chop chop.

"She would start with hot oil and throw in mustard seeds. You heard them popping. She never had a recipe; she just knew the ideal mix. Onion and turmeric and ginger, peppercorns and garlic, coriander, curry leaves—all fresh. It was very, very fragrant. And the *appam*, the rice pancakes. And the chicken that would just slide off the bone when she was finished. We sat for dinner at this faux marble table with metal around the edges, and there were six or seven huge bowls of vegetables and things. We ate with our hands, and I would suck every bone clean. The *appam* would soak up the curry perfectly. She was a whole-spice cooker, so an extra part of the texture, an extra part of the experience, was biting unaware into a mustard seed or a curry leaf—I loved that.

"The only problem—and it wasn't enough to change how much I looked forward to her curry—was that the smell, the smokiness, stayed on our clothes. I was hypersensitive to that. I would get upset when we walked out of the house with our clothes smelling of our food. I thought I could protect myself if I didn't smell like that, and on my way to school I would throw my coat, my sweater, into the bushes, hoping they would be there for me to pick up later. It didn't work—the smell clung to me; there was no way to get rid of the fragrance of our cooking."

A voice began to draw his attention to the way his mother watched him eat. It introduced a question—Why was she doing that?—and one evening gave an answer. With the curry she had just served, she was trying to poison him. The voice warned against taking a bite. His mother stared at him, stared at his plate, his untouched *appam*, his untouched curry. She asked him about it; she interrogated him; he refused to explain. His father and sister finished their meals and left. His mother's words grew sharper.

She said that he could not leave the table until he ate, and he stood, flipped over his plate, flung his chair.

Evening after evening, she watched, and he kept up his resistance. He would not let her kill him. Her gaze, even when they weren't fighting, inflicted a physical sensation of being stabbed. At night, in bed, he had liked to listen to his parents as they talked in Malayalam, their tones and the language soothing, their words muffled and beyond meaning, but now they were plotting against him, hoping to be rid of him, and his voices urged him to stay vigilant, to decipher their words and discern their plans. Shadows leapt around his bedroom, misshapen and horrific, and a woman sprawled on top of him, a white woman with high cheekbones and white hair, but with a darkness, a shadow-woman, in flowing robes, struggling to get inside him as he struggled to scream, all while he was wide awake. He told himself he was concocting her but couldn't unsee her, until she was replaced by a demon with a hypodermic in its hand, its injection lethal and linked, he was sure, to his parents, on the evidence that their arms bore vaccination scars.

The scheme to poison his food; the woman both shadowy and white; the demon with its needle; his fearful, protective voices—all continued. "I went," he said, "from being frightened at school, to frightened on the walk between my house and school, to frightened in my house. It was like it seeped inward."

As a teenager, he quit showing up for dinner, and his parents quit insisting. He lost weight and, though he was a fair athlete and a skilled soccer player on a travel team, became the target of yet more abuse. "I drank. I found kids who helped me use drugs. That helped me temporarily." But the ridicule that might or might not have been real, the nightmares that weren't nightmares, the voices, the certainty that his parents saw him as a taint to be purged— these were stronger than the narcotics. Still, he kept using and started dealing. He had, at fifteen, a few nascent friends, football players, because of the drugs and booze he supplied.

One afternoon, in early December, after a pep rally at school, he planned to deliver an offering of weed and alcohol to the football players to solidify their friendship. But as he walked to meet his dealer, a pack of kids mocked and threatened from across the avenue. "To this day," he recounted, "I don't know if they were there or not, but they were yelling, wanting to fight me, tracking me for blocks, past the town hall, past the police station, the library. I got to a traffic light, and I decided. I was resolved. *I can't do this anymore. This is it. This is over.* And everything quieted down. I'd had these thoughts before, but this was the first time I knew it was going to happen."

That night, he smoked enough weed to keep himself from throwing up, and guzzled enough liquor and smoked enough freebase to kill himself. He came to the next day with his wrists strapped to a hospital bed.

Since then, thirty-four years had gone by. "Today is the anniversary of the night, and I'm close to those feelings," he said during one of our conversations, as we sat in the office that Dixon's New York State–funded organization, aimed at adolescents and young adults with psychosis, rented for him. "That my parents were trying to kill me. That I didn't belong anywhere on earth."

All he related seemed at odds with the man sitting across from me. The office was stark, with only an airshaft window—it was in an apartment above a subway station, a kind of random satellite to Dixon's headquarters, a few blocks away, in a new and mammoth Columbia building on the bank of the Hudson River—but Chacku's skin had a glow, his smile an ease, and his chest a breadth and sturdiness that announced radiant health and resilient youth even in the semi-darkness. Such were his surfaces. He said, "I experience the world as a suspect perception, and I have to work to figure out what your perceived reality is, in order to fit into it and operate within it, in order to interact with your interpretation. I'm hyperconscious of your interpretation. Do I know there's a consensual reality? Yes, to some extent. There's a recorder on the table recording our voices. There is a table and chairs. But sometimes I'll have to ask someone I trust, 'Did you see that? Did you hear that? Is that person talking?' I'm under a lot of stress today, and I'm hearing voices intensely."

At the hospital, Chacku's parents resisted the recommendation to put him on an antipsychotic. It was an alien idea to them, alien to their culture. There was pressure, the perception that they were irresponsible or misguided or worse. They began to debate moving back to Kerala, their state in India. As a child, on a visit to his father's village, Chacku had been instantly at home, climbing onto the back of an elephant that was clearing the fields, the animal pausing to receive its small, reverent rider, who scrambled onto the bent, scaly knee and clutched the stupendous ear and tugged himself high. A village girl befriended him. They played soccer with a ball she fashioned from coconut leaves.

The family didn't move back; instead, Chacku's father met a man whose son had attempted suicide and been hospitalized, a father who had taken a leave from his job to create a space, in an old horse barn, where teens like his child could find refuge. Chacku's father took Chacku there. "It was skinheads around a pool table. I was scrawny, with long hair I refused to cut. I thought, *Here we go*. But one of them asked if I wanted to take the next shot. He joked that maybe I would get a haircut someday, and somehow I didn't hear it as an insult and a challenge to fight."

Besides the barn, Chacku spent all the time he could at a kill shelter. As comfortable as he let himself feel among the skinheads, he didn't develop a deep trust in anyone; people remained covertly menacing. His mother's curry "tasted like hate or disdain or the wish that I was gone." But the dogs and cats who awaited adoption, if they were lucky, and, in a back area of cages, the pit bulls and rottweilers who had bitten someone or been used for fighting and gambling, and who awaited adjudication and likely euthanizing—among them, he trusted. He adopted a German shepherd, an avid hunter of groundhogs and muskrats, and then took in an untrained dog he found through an ad in the newspaper classifieds. The dog had ripped flesh off a man's arm and was imprisoned in a basement. Chacku's German shepherd tamed it, and the two dogs became Chacku's closest family.

Through one of the parents involved with the barn, he was asked to speak, when he was sixteen, to younger at-risk kids in a locked mental health facility, most of them kids of color. He told some of his story and listened to whatever they were willing to share. He was invited back and, over the next years, invited more widely. But he knew better than to let his guard down; he kept his voices to himself. At twenty, he attended a talk given by Lionel Aldridge, a legendary defensive lineman and two-time Super Bowl winner with the Green Bay Packers, who, after his retirement, talked publicly about his battle with schizophrenia, about the accusatory voices he heard and the belligerent, conspiratorial surveillance he sensed in the TV cameras pointed at him while he worked as a commentator during games, about his spiral into homelessness.

Chacku rushed to the front to get in line after Aldridge's speech. When his turn came, he knew he had only a minute; he blurted out his question. He said that he'd been hearing about the cyclical nature of his illness, and he asked the ex-lineman if it was true that the harshest of his voices and the hardest times would return over and over forever, cyclically, unchangingly, for his entire life. "No," Aldridge told him. "That's not right. It's not like that. You're always growing." Aldridge explained that from the outside the cycles might seem nothing more than repetitive, but that from the inside there would be progress: around, outward, upward. "In my mind, as I left," Chacku remembered, "I wasn't seeing recurrent cycles, I was seeing expanding spheres."

He immersed himself in Hindu texts. His parents weren't Hindu; they had been raised in a mystical Christianity dating back to Saint Thomas's gnostic teachings in southern India in the decades after Jesus' death. It emphasized koan-like sayings and quests of spiritual enlightenment. Chacku's father had once told his son that he wanted him to be able to pray with all people, and now Chacku studied the Bhagavad Gita, the seven hundred Hindu verses composed perhaps fourteen centuries ago. He practiced yoga and performed *puja*—the rite of songs and recitations honoring Hindu deities—multiple times each day. He sounded a mantra, aloud or silently, filling the air each morning with sacred vibrations. All of this granted his mind a degree of rest. He learned from a tantric community and its guru, extending the wisdom Aldridge had given him. "I became able, as I meditated," he said, "to see swirls of light moving through my chakras." His expanding spheres were luminous.

He had deities tattooed onto his arms, to focus on whenever needed, along with a tattoo of his childhood elephant. "I felt so connected when he gave me his knee and his ear. I felt his bristly hair. I felt his permission. Of course, now I wonder about his oppression, but then there was so much belonging, everything I couldn't find in the States." He beckoned memories of the river that ran through his father's village. His uncle warned of snakes, but Chacku dove in. "I kept my head underwater and my eyes open. The sun was filtering through. A fish with a long nose bumped my legs—bumping, bumping. It was the best feeling in the world."

Bodies of water became, in adulthood, the medium of what he called "a delocalized consciousness," an antidote to the solitary and harrowing, the self-protective and self-predatory contents of his mind. Evocations of oceans, submerging him in their waves while he meditated, allowed him to "leave the local behind, expand into the greater, find a way for union." His religious practice let him escape his extreme subjectivity. He began, in his thirties, to glean love in his mother's gaze. The oceanic was a kind of cure for his disconnection.

He thought often, he told me, about whether, if a miraculously improved antipsychotic existed, he would take it, or would have when younger. "My experience is so rich," he said, "I wouldn't trade it for anything." Again, he invoked water, the image of a free diver descending without oxygen and using a rope to keep his bearings and guide himself down to unfathomable depths, places others couldn't reach. For him, the enabling rope was woven of his voices and visions. He spoke, too, of having a keen awareness of the singularity of others, the solitude of others, a sensitivity that was paired, paradoxically, with the feeling of being universally joined.

None of this meant that he spent his days in a state of watery bliss. He managed himself, maintained his vigilance. He took care not to talk back aloud to his voices or to heed their admonitions—less unrelenting than years ago but never gone—that he was under assault or facing unjust opposition or being ignored and that he should flee or lash out. In this way, he built a career, training peer specialists and serving on the advisory boards of mental health groups and eventually running a small organization in Rochester. His experience with psychosis—his "lived experience," as it was newly designated and seemingly esteemed—had become an asset, a credential, a badge that benefited his employers. Yet for the comfort of his colleagues, he made sure to give the impression that his voices were scarcely part of his present life, that his experience was quarantined in the past.

Chacku gave speeches to groups aligned against the authority of biological psychiatry, speeches that expressed some of what he'd lived through and the lessons he'd drawn, but shortly before we first met, he had taken a position with Dixon's OnTrackNY, whose reputation was progressive but whose approach stuck close to convention. OnTrackNY targeted people in their late teens and twenties who were starting to deal with psychosis, and its core reform was its proclaimed "person-centered" care, a commitment to include the ideas and wishes of the afflicted, as well as the paradigms of professionals, in planning out treatment. Dixon, like Goff, was nothing if not reflective. The bleak life of her older brother, who had been diagnosed with schizophrenia as a young man, occupied her thoughts. She had spoken to me about "according personhood" to patients and about the paltriness of knowledge about psychosis. "We don't know who needs what," she said. "The illness is so heterogenous. We're so handicapped."

But her program, which coordinated counseling on employment, education, and social skills along with coaching for families, clung to the biomedical. Antipsychotics were central and essential. The treatment manual advised: "Team members including the prescriber and nurse should work with individuals who decide not to take antipsychotics . . . to regularly revisit the decision." The manual acknowledged that many patients abandoned their medication but attributed this mostly to patients' lack of insight into themselves and their medical needs: "Anywhere from 30–60% of individuals will be inconsistent with medication or stop medication altogether. . . . If an individual is ambivalent about using antipsychotic medication, the handout 'Finding Personal Motivation to Use Medication' may allow them to identify how antipsychotic medication can help them to realize their own goals. . . . Long-acting injectable antipsychotics (LAIs) are a potential pharmacological approach that may enhance adherence."

Reading the manual, which gave such minimal weight to the individual's own perspective, it was hard to believe that Chacku felt he belonged at Dixon's organization, especially because his work involved reaching out to young people and bringing them into the program, and because the program aimed to medicate people undergoing an initial episode of psychosis, without waiting to find out whose symptoms might subside on their own and never return. "Lifelong medication" was the stated goal. "Person-centered" could seem a professional self-delusion. But Chacku said that he found reassurance in Dixon's "humility" and solace in her willingness to listen. He would proceed slowly. "I'm interested in appealing to their desire to learn," he said about his colleagues, even as he felt the limits of that desire and an underlying defensiveness against much that he hoped to impart.

"How close do you think we are, as a society," I asked, "to the kind of significant change you would like to see?"

He laughed. Then, in answering, he referred to a stand of elms in a park not far from the apartment he shared with his wife and eleven-year-old daughter. He'd told me that he often sought out this set of magnificent trees when his voices felt unmanageable. A beneficent voice had helped him to find the spot, "like being guided by one of those water-finding sticks." And just a few days ago, he'd overheard his daughter on the phone with one of her friends. "She said, 'My dad was talking to the trees.' And then she goes, 'Yeah, he talks to trees.' That's her, understanding who I am in a way that's not pathologized. That's my significant change. Are we close to that as a society? No. But it doesn't mean we can't get there. I didn't expect statues to get pulled down this year either."

RACHEL YEHUDA WAS Nestler's colleague and the director of Mount Sinai's Center for Psychedelic Psychotherapy and Trauma Research. She also oversaw research with the department of veterans affairs. After David's indifferent experience with psilocybin, I had contacted her for an overview of current studies on psychedelics and mental health, beyond Griffiths's work at Hopkins, but our conversations ranged widely.

Yehuda's long career had been defined by the physiology of the brain; her latest paper looked at "glucocorticoid-induced transcriptomics in human neurons." Like Nestler, she was drawn to the mechanisms within neurons that made some people resilient against stress and others more vulnerable, and this paper described an experiment comparing combat veterans with and without post-traumatic stress disorder, PTSD. She had been able to illuminate—indirectly, suggestively—pertinent variations within cells. It was formidable work, but she had grown dissatisfied with her accomplishments. She doubted that her discoveries would ever be helpful to human beings. "The science of what's broken in the brain may not be the science of healing," she said; experiments like the one she'd just written about, pointing to a molecular factor potentially involved in PTSD, would all too likely prove useless in treating someone's anxiety and emotional pain. "We're fucking it up in mental health," she told me.

Lately, she'd taken a detour into territory that made her uneasy and that she sometimes categorized as "woo-woo." The detour had begun at Burning Man, the desert festival of electronic music, hallucinogens, and sculpture set aflame, which she'd first attended in her fifties, a few years before we talked. She hadn't taken psychedelics at the festival, not that summer or at the Burning Mans she'd been to since, but she found herself ministering to—mostly just listening attentively to—young people whose drug-fueled journeys led them to tumultuous places. "I'm a person," she explained, "who will do anything to avoid small talk." Here she was enveloped in small talk's opposite, embraced by the openness of the festival pilgrims who surrounded her, and inspired, two or three years after her introduction to Burning Man, to have a psychedelic experience of her own.

She clarified that she'd arranged to do this legally, then said that she'd been able to meet her dead father and contemplate, viscerally, a biblical passage that was linked to him in her mind:

Moses asking to see God's face and God refusing yet revealing a fragment of Himself. She choked up as she related this and recalled the meaning that emerged from it, that she was not alone in a family crisis she was going through; that here was an answer to "Who is going to help me carry this burden?"; that "my ancestors are here with me"; that "you're part of something bigger."

Yehuda spoke about her role in ongoing studies investigating whether and how therapist-guided psychedelic sessions might treat severe depression or PTSD by fostering the sense of connection that she'd reached during her own session and that she'd cultivated, without the help of more hallucinogens, ever since. One of our conversations followed a *New York Times* story that covered two experimental trials of psychedelic treatments—neither involving Yehuda—and that ran under the headline "The Psychedelic Revolution Is Coming." But it seemed that journalism was repeating the past in rushing to announce psychotropic breakthroughs and feed our hunger for chemical cures. Of the two published papers that detailed the trials, only one contained positive results. The other concluded: "This trial did not show a significant difference in antidepressant effects" between the psychedelic and the SSRI that was used for comparison. The *Times* didn't acknowledge this statement; it was as if the journalist preferred not to reckon with this outcome. And the story hardly noted the many hours of specialized therapy that were woven into the one successful trial and that helped subjects to process, in the words of the therapists' training manual, psychedelic-induced "perceptions that are felt to extend well beyond the usual sense of self," "enhanced connection to their own humanity and the surrounding environment," and "feelings of oneness." Perhaps this sort of thing was just too woo-woo for the newspaper, but it was a key part of the trial's method.

"It is critical not to lose sight of the fact that these compounds are occasioning an opportunity for insight," Yehuda said. "They are not magic pills that automatically reset the psyche. Psychedelics," she paraphrased Stanislav Groff, a Czech psychiatrist and advocate of spiritual approaches to therapy, "can be to psychology what the microscope is to biology and what the telescope is to astronomy—the psychedelic is a tool that permits the patient to see something that cannot be seen or appreciated under normal circumstances. And once that insight is made, it is important to be full of curiosity about what it is and what it means and how it can make a difference. That is best done with a trained therapist who can facilitate the process."

No matter how long-term research into psychedelic treatments turned out, no matter whether psychedelics would have a role in mainstream psychiatry, Yehuda was certain that it was time for a new paradigm, a pivot away from the prevailing mode of "suppressing symptoms" and toward—she apologized for the woo-woo that was on the way—"an expansiveness that can *hold* symptoms."

She asked, "What does it mean to heal from a problem of mental health? Is it the same as healing from a broken bone or a physical disease? Or is it a completely different thing?" Did we need to "make peace" with our fear and pain, to welcome them, to grant them an integral place within ourselves? "'Hello, darkness, my old friend,'" she quoted Paul Simon's lyric, straining to encapsulate her ideas. "Does any of this make sense?" she asked me.

We were a long way from the language of "glucocorticoid-induced transcriptomics in human neurons." I thought, as we went on talking, about Chacku's expansion—and dissolving—into bodies of water; about Nestler's brief mention of "the belief in something bigger than yourself"; about aspects of my brother's story—and Caroline's—that I will come to; of times, as the leadenness and weakness of my arms overwhelms me, when I am wise enough—which is not always, by any means—to walk outside and find a place where I can lie back or even just tip my neck far back, where I can stare straight up into the branches of a tree, stare for some fair length of time, and watch the movement of those limbs in the wind or study their intricate stillness if the air is still, and allow myself to feel—despite all arguments to the contrary and despite the secular adamance that dominates my world as a journalist—that the movement or stillness of those branches is God speaking, or the Universe speaking, saying to this speck-like being, *You are part of Me.*

ELEVEN

D o what is asked of you," the voice said.

It was three weeks after that July Fourth, after her plan to kill herself was diverted by a man in distress and a display of fireworks. The voice was responding to the offer of a dismally paying part-time job eight hundred miles away in Massachusetts. Her favorite co-worker at the Asheville hospital had told her about the opportunity, not knowing how minimal the hours and poor the compensation would turn out to be. And now her prospective boss at the Massachusetts organization was advising her, over the phone, to say no, to stay where she was, making a decent salary, and not to uproot herself for something that would give her so little and that came with no guarantee of anything more.

But the voice was definitive. "Do what is asked of you," it repeated, and Caroline told the woman on the phone that nothing could dissuade her, that she was coming.

She packed up her uncle's beleaguered sedan and drove north in yearning and anger: anger at a system that had divided her so fundamentally and profoundly from her old housemate at the group home, that had made it impossible for Margaret to confide her suicidal feelings and impossible for Caroline to hear them, that so prioritized the management of risk and the prevention of suicide that it prevented communication and connection, the very things—as she had listened to the rape victim's Fourth of July story and his desperation to set that history right this year, and as they had perched side by side on the van's roof and watched the lavender jellyfish launched into the air—the very things that had saved her from taking her own life.

She got as far as Allentown, Pennsylvania, slept at an Econo Lodge, and kept driving the next morning, ruminating, veering between reflection and outrage, as she hurtled up the interstate. At the Houston hospital, it had seemed that the staff had focused more than anything on keeping death at bay, on recognizing and removing the most improbable means, like her pinky-thin, three-tenths-of-an-ounce tube of mascara, on banishing topics at group sessions that might lead anyone to dwell too darkly, on cautioning against forming friendships, as if any bond between two troubled minds could only multiply the trouble and risk for each. Best to keep people chemically subdued and interactions carefully regulated. The exception had been the smoking cage, where so much of the conversation and bonding was around the subject that was most taboo, suicide, and where she felt least alone and most appreciated, farthest from suicidal and closest to happy.

And the hospital was just one symbol of the entire psychiatric system. How could this be the way? How could this system, which she knew so well from decades as a patient and years as a peer specialist, and which before anything else reduced you to whether or not you thought about killing yourself and whether or not you thought about how, and which, depending on your replies, raised such deafening alarms that you as a person of singular nuance and fear and complication and loneliness and insight and need could no longer be heard and were replaced by a creature to be kept alive—how could this be the best way to help human beings? It was, she felt, absurd, all the more so, she railed in the dented, semi-jalopy of a sedan, because the professionals posed these first-principle, crudely binary, soul-crushing questions and believed in them and acted on them despite knowing what study after study had shown: that suicide risk could not be measured, that the answers to the questions were too often lies and even when candid carried too many different meanings.

In this state of highly informed fury twenty-some days after forgoing her own planned and prepared overdose and asphyxiation, Caroline arrived in Holyoke, Massachusetts, and presented herself for a job that scarcely existed. She presented herself well. She had become adept at controlling internal turmoil and converting it to outward determination, to directed motion. She credited this to roller derby, to the feeling of mayhem, in her own mind and in the energies of her teammates, being marshaled and deployed.

"After working within the system," she told me one day in Holyoke, "I came up here with so much to prove. I needed to prove who I was, and that the system could be dislodged, and that I was ready for whatever it took. And as soon as I got here, I was at one of the centers, and the toilet got clogged. And I double-bagged my hand, and I dislodged whatever was in there, because I needed to know about myself, I am here for this shit, I am here to find out what's possible."

The job was with a fledgling nonprofit, the Western Massachusetts Recovery Learning Community, headquartered in Holyoke, a withering city that had once been the home of the biggest paper mill in the country and was now one of the poorest places in the state. Its downtown was lined with empty storefronts and shops whose owners seemed to have given up long ago on arranging their window cases to entice customers. The windows were bleary, their displays sparse and forlorn, the few people on the streets somehow fragile, as if only society's marginal citizens would appear here—as if to walk these desolate blocks was to be exposed as weak and unwanted.

When Caroline showed up, the organization, run and staffed by people on their own psychiatric journeys, was days from opening what was known as a peer respite, a house, as the community's website said, for those who were "walking around this world pretending that everything is okay until they just can't pretend anymore." It was one of a dozen or so homes like it around the country, a peer-operated option to checking yourself into the hospital that was akin to Lichtenberg's Soteria houses in spirit, though it had only the wherewithal to put people up for a week while they sought a path forward. The house made a point of welcoming the suicidal, and the website critiqued the idea that "the best thing for people who are thinking about killing themselves is to be on a locked unit where they can't cause themselves harm, not a peer respite where they're free to come and go."

Caroline's vaguely outlined job involved pitching in as a peer counselor at the respite and conducting Hearing Voices meetings in and around Holyoke. The Hearing Voices Network in the United States was part of an international movement, a constellation of support groups, from New Zealand to Bosnia to Greece to Norway, whose participants heard and saw what most would say was unreal. HVN had its origins, in the mid-1980s, in the relationship between a Dutch patient, Patsy Hage, schizophrenic and suicidal, and her psychiatrist, Marius Romme, and in Hage's insistence that Romme pay some attention to the content of her voices instead of dismissing what they said as meaningless. One version of the story had Hage demanding to know how Romme could believe in an unheard and unseen God while refusing to entertain the reality of the voices she actually heard. In any case, Hage and Romme wound up being featured on a Dutch TV show and receiving hundreds of responses from voice hearers, a third of whom, Romme learned as he began to study them, coped with their hallucinations without medication and without having their lives derailed. Some probably weren't diagnosable as psychotic; they spoke occasionally to a family member who had died. But plenty could be categorized as ill yet managed their disorder on their own.

Romme catalogued their methods in a mainstream psychiatric journal, writing that "it seems very important for individuals . . . to attribute meaning to the voices"; that it was possible for them to reach "peaceful accommodation and acceptance" of their voices; that practitioners should "accept the patient's experience of the voices"; that they should "consider helping the individual communicate with the voices"; that "the explanation offered by biological psychiatry" was "not very helpful in coping with the voices, because it places the phenomenon beyond one's grasp"; that practitioners should "stimulate the patient to meet other people with similar experiences"; and that "for most psychiatrists, these steps will require an enlargement of one's perspective."

The paper was widely ignored, but scattered HVN groups cropped up quickly in the Netherlands and Britain, more slowly in the rest of Europe as well as Australia and Canada, Japan and Kenya, and slowest of all in the United States, where some of the first groups were founded by the Recovery Learning Community in the Holyoke area not long before Caroline got there in 2012.

Besides helping out at the respite and running a few HVN meetings, Caroline had a third role in another barely hatched project. She was supposed to facilitate a suicide prevention group that would substitute pure listening for the focus on risk, the fear of liability, the reflex of intervention, and the predominant assumption that those who spoke of imminent intentions should be sent into a system where they could be kept from death. She was charged with getting people to talk freely about wanting to kill themselves, without the worry that if they did, they would be locked up and most likely heavily drugged. She did this at a center several miles from Holyoke, one of four run by the community. It was a den of deteriorating furniture with a glorious mural occupying all of one wall, a medley of yellows and browns and blues, of lovers and a galloping horse and a playful caterpillar and a Puerto Rican flag and President Obama, painted by the disoriented and discarded. It was here that Caroline volunteered to clear the toilet, to let everyone know, including herself, that she was committed.

I FIRST MET Caroline in the months before the pandemic, in Holyoke's lone café, a bare and cavernous space, empty except for us and the owner. Caroline wore a red cap, its thick wool doubled over and tugged down close to her eyebrows, and wool gloves with half fingers. It was as if she wanted to be in the protective gear, the helmet and pads, she'd worn for roller derby. She seemed to anticipate that the world—or I—might try to knock her down. Or she may only have been chilled. The cavern was underheated. She sat slightly hunched in a booth, dark hair past her shoulders, skin pale, chin delicate, eyes with a gentle gleam, and said that she'd been talking with her voices about her talking with me.

"I've told them that you're not here to hurt them," she said. She explained that the voice she'd known the longest had been very wary but that he was speaking to her now, expressing a preliminary trust. Her voices spoke to her through most of our time together that afternoon and evening. She could hear me over them, and could think and formulate her own words over them, except for one moment when she asked me to repeat myself, because her voices had drowned me out.

I had a minor idea of what she meant. At her suggestion, I'd been to a three-day HVN training workshop in the town of San Marcos, in southern Texas, halfway between Austin and San Antonio. The training was for people who might want to start an HVN support group. It had been seven years since Caroline had begun working with the community. Scant though the organization was—its Holyoke offices consisted of three rooms—it had become an HVN hub. Much of HVN's spread throughout the country was being engineered out of Holyoke, and much of it was, by now, being led by Caroline, who was more than fully employed, along with the community's director, Sera Davidow, who'd accumulated a half dozen diagnoses in her teens, been hospitalized for the first of two times at twenty-two, and recalled, "I learned not to be honest; *do not tell* above all else." She relied on C. S. Lewis for wisdom: "Of all the tyrannies, a tyranny sincerely exercised for the good of its victims may be the most oppressive."

In less than a decade, HVN had grown to about one hundred and twenty groups around the country. This didn't come close to satisfying Caroline. For one thing, the UK had one hundred and eighty with a population one-fifth the size. For another, most of America had none, with the groups concentrated in the Northeast or on the West Coast. There were only two in all of Texas, and trainings like this one, in a conference room with oatmeal-colored walls and a leaky ceiling at a Holiday Inn on an access road of strip malls, were a way to plant seeds. Thirty people sat around a formation of fold-out tables and listened to two presenters, who'd been sent from Holyoke. They introduced HVN's mission and principles, which were all about sharing and validating the subjectively real, softening the most extreme solitude, and perhaps making sense of what much of psychiatry took to be chemical, random, and senseless—the way Chacku, a participant in HVN groups, had done in tracing his hallucinations and delusions to the dislocation and xenophobia that had surrounded him as a child.

The trainees in San Marcos were a mix of social workers and psychosis sufferers. A few may have been both. The counselors had been sent by mental health nonprofits and local school districts. There seemed to be widespread curiosity about what HVN groups might provide, generated by the community's webinars and online newsletter and by a TED talk, with five million views, given by a British woman who told of being "diagnosed, drugged, and discarded" as schizophrenic when she was a university student, and who'd "attempted to drill a hole in my head" in order to extract her voices, but who'd learned to interpret them as having "something to communicate to me about my emotions, particularly emotions which were remote and inaccessible," and who'd gone on to spectacular academic success and a career as a psychologist. And in San Marcos, as everyone split into small circles to practice what they'd been taught, with one member of each circle assigned to play someone in crisis, there seemed to be a willingness to trust that when someone shared what she'd been hearing, it was better to ask for more detail or to ask others in the circle if they'd heard similar yelling or similar whispers, than to remind the person that her voices weren't real or to give advice about medication or to rush into asking if the voices were telling her to hurt herself.

The impulse to fix was to be kept in check. The need to save could do more harm than good, more damage than repair. All that was required was to listen. One of the presenters spoke about how being sexually abused as a child might have given rise to her hallucinations and said that she despised the concept of triggering, despaired over the professional fear that talk of certain subjects was more than she could handle. "It implies that if I have a big feeling I will go off. It implies that I'm a glass bomb. But I live my life with big feelings."

During a break, I had lunch with a school counselor who was moved but unconvinced. She couldn't imagine starting an HVN group in one of her middle or high schools. What was she supposed to call it to lend some privacy to the kids and protect them from being hounded by their classmates? And how was she supposed to deal with the obligation to report their problems to their parents?

"If they talk about hearing God—around here, no, I don't have to report that. But signs of psychosis?"

An African-American man who'd been a prison guard and now worked with ex-cons was more inspired. "Come back in two weeks, and you'll see I've started a group," he told me. "A few of my guys, believe me, they are going to benefit. They need to exchange some strategies." This was one of the HVN tenets, that there were techniques for tending to your voices, that one value of the groups was the trading of homespun methods, that your relationship with your voices—more than the rare result of medicating them out of existence—could liberate you, balance you, bring you closer to whatever was meant by mental health. The trainees were given three-ring binders full of guidance, addressed mostly to voice-hearers themselves, about how to relate: to understand that your voices often have meaning, sometimes indirect or metaphorical, in the context of your life, particularly in the context of trauma; that it can be helpful to recognize your voices' needs but also to negotiate specific times for checking in with them while letting them know they will not get attention whenever they clamor for it; that "thriving with voices" is most likely when they are "seen more as belonging to the individual and have a message."

After dividing into practice circles, the trainees split off again, into threes, for an exercise. I'd taken part in the circles, and I took part now, in what turned out to be, for me, an unforgettable segment, though the exercise was brief. I sat for a job interview. Across from me, a trainee asked run-of-the-mill questions about my work experience, abilities as a team player, hobbies. Another trainee stood beside and above me, with a long tube of wrapping paper running from her lips to my ear. She spoke into the tube. She was a military veteran who'd recovered from a brain injury after she was knocked unconscious in an accident on base. She worked now as a peer specialist, sometimes with voice-hearers. She'd said she made a point of trying to learn from them. And she'd had an aunt she was close with, diagnosed with schizophrenia, who'd taken medication but was never free of voices; she'd taken her own life. So this woman, who wore a red sweatshirt with the hood loosely around her head, and who sent her voice into my ear via the cardboard tube, knew what she was doing.

"Careful what you say," she, my voice, warned as I went through the interview. "Careful what you say about your background." The voice, amplified and concentrated as it traveled through the tube, burrowed and vibrated in my inner ear. "Careful what you say about your background," it cautioned again. I couldn't ignore it. I couldn't think coherently. I was tempted to turn toward it, to tell it to quit, though I was fully aware that the interview was just a game. The voice kept on with its vague insinuation about my past. I was drawn into trying to figure out what it was referring to. The insinuation was unspecified but didn't feel unspecific. I was more than distracted. I couldn't come up with suitably innocuous answers to the interviewer's benign questions. I couldn't produce answers at all, for fear that what flew from my mouth would be a violent "Shut up" aimed at the voice. I sputtered into silence. The interview came to a halt. I said, "I give up."

This is what I thought of as I sat with Caroline in the café and she asked me to repeat myself because for a second or two her voices had grown too loud. This was her battle. This was her life. She did all the prescribed things, existed in relationship with her voices rather than in opposition to them; she had overseen the crafting of the training curriculum and helped to write the prescriptions in the three-ring binders; yet her burden could only be reduced, not removed. This was her extraordinary trial. This was what she bore, even as she sipped tea in a booth in our shared world, and chatted amiably, charmingly, with the café's owner, and spoke keenly and lovingly with me about literature and poetically about roller derby, and led us into the December wind of downtown Holyoke and along the empty sidewalk to her office.

There, she had a map of the country on the wall, dotted with colored pins. "I sometimes feel like a general mapping the revolution," she said. "But this is also here so our staff and volunteers can see how hard they've worked." Blue pins, from Augusta, Maine, to Eureka, California, marked places where the community had orchestrated training like the workshop I'd been to. Groups that had developed were designated one of two levels: the lower, founded and run by any sort of clinician, like the social workers in San Marcos; and the higher, led by someone who heard or saw, with no clinician allowed in, unless he was afflicted himself. The ideal was to dispense with outside expertise, which was inherently limited in insight, though, for the time being, in order to make progress, there was sometimes no avoiding the practitioners.

Red pins stood for cities and towns where Caroline or the community's team had conducted trainings for would-be facilitators of her suicide prevention groups—the program that had been half-conceived and consisted of a single group when she'd arrived in Holyoke. There were now thirty regular meetings across the country, from Boston to Denver. Over the next two years, there would be seeds planted, in the form of trainings given, some by Zoom, in dozens more spots around the United States. During the previous year, Brazil's mental health agency had brought her to that country, where she'd spoken to crowds of clergy and clinicians, school administrators and law enforcement officials, as well as the diagnosed in São Paulo and Salvador and Vitória. Mercy Care International, a worldwide charity, had flown her to Australia to speak in Sydney, Melbourne, and Perth. Speeches were upcoming in Indianapolis and Portland, Oregon. Something was happening, something that, in large part, Caroline had created.

The suicide prevention program had a bland name, Alternatives to Suicide, and a transformative agenda. As with HVN, the hope was to keep clinicians out of the room, and the requirement, regardless of who was leading the meeting, was to let people talk freely—and confidentially—about all that was tormenting them. The foundational pact was that no one would be reported, not to any hotline, not to the police, not to any practitioner, no matter what they said they longed to do or were on the brink of doing.

A bedrock faith lay within the crusade that the community and Caroline were waging. It was in the respite's credo that it could be better for a despondent and suicidal person to stay in an unlocked, peer-run house than on a high-security, professionally governed ward. It was in HVN's belief that describing voices and visions in detail, that filling a room with phantasms, would not infuse them with more vivid life or grant them more intransigent power, but would, partly by lifting the pressure of secrecy and diminishing the feeling of deviance, loosen their hold. And it was in a key principle of Alternatives to Suicide: that as long as you were talking about killing yourself, and felt you were being listened to and understood, you were much less likely to end your life, that it was the loss of connection that led to death.

The reaction to all of this could so easily be terror. If my loved one was suicidal, didn't I want him in a hospital with his shoelaces and all other means taken away and an orderly to monitor his every minute? Didn't I want him to be *safe*? In a crisis, wasn't that the all-consuming goal? Couldn't all else wait? And if my loved one was psychotic, didn't I long for her visitations to be banished, her difference silenced and erased, her normalness restored, time overruled and turned backward to whenever she was or seemed to be like everyone else; and didn't I want her strangeness subdued to any degree possible by the magic of medicine; and did I want her anywhere near some dingy room of sofas with torn upholstery and chairs with splintered legs and diseased people evoking the very kinds of visitations I wished expunged from her brain, evoking them as if they existed and never being corrected and told they absolutely didn't—did I want her to step into, to surrender herself to, to sacrifice herself within such a realm of horrors, such a chamber of the lost? And if she attended a meeting and said that she planned to kill herself, didn't I want someone to dial 911?

Terror was what Caroline's crusade would need somehow to conquer, a terror whose forms I knew well enough from my parents—and from the dozens of stories I listened to and learned from, stories submerged just beneath the surface of these pages—a terror that arose at an Alternatives to Suicide training I took part in, again for people interested in founding groups, again in the conference room of a hotel on a road of strip malls, this time a half hour from Holyoke, with Caroline teaching.

It was during a lull in the pandemic, and she had decided to hold the workshop in person, with everyone masked and spread more than six feet apart, because she could hardly bear to lead another session by Zoom on such excruciatingly intimate subjects as hearing voices or wishing for death. She wore jeans and a letterman-style jacket in the heavily air-conditioned room, but no cap, no half-fingered gloves; she looked and sounded at ease as she clicked through slides on the screen behind her—"Rather than assume the need to contain or solve someone else's issues, we listen with deep and true interest"—and posed questions of her audience. She asked whether we agreed with the statement "I believe that people should be stopped from killing themselves by any means necessary."

A middle-aged woman raised her hand. She said she didn't know the right answer. She was there because her husband had committed suicide in their garage. "A few days before I pulled him from the car," she said, "I told him I needed him for my companion. I told him I needed him to embrace life. I told him I needed him not to kill himself. I told him to get a job bartending in a strip club, because I needed him to live."

The raw helplessness in her story brought silence in its wake, but Caroline's question was provocative, and the silence didn't last. People talked about the problem, and I thought about the policy of the country's most-called suicide prevention hotline. It advertised confidentiality but covertly scored risk and traced calls and, each year, without permission, dispatched police cars and ambulances to the doors of tens of thousands.

Later, Caroline sent us outside, in groups of four, to discuss a related question: If we were leading an Alternatives to Suicide meeting, was there anything a participant could say that would cause us to break with the pledge of privacy and call 911? My group found some benches on a grass island in the middle of the hotel's parking lot. All three of the people with me had spent time on psych wards, and the workshop had already made the case for never breaching a participant's confidentiality or impinging on her autonomy. Caroline had argued, "We don't hear people say, 'I stopped thinking suicidal thoughts when they took away my shoelaces and belt at the hospital,' but we do hear people say, 'I stopped *talking* about my thoughts.'" And she had said, "When I'm controlling, I'm not connecting." So on the parking lot island, our group was primed to declare that we would never betray our pledge.

But then a man asked, "What if someone says they're going to kill someone else?" The conversation took a sharp turn. Swiftly the three agreed that talk of murder would be a different category, that the only thing would be to call for help.

Back inside the conference room, everyone reported on their conversations. Another group had taken the same sharp turn and come to the same conclusion. Nearly all the trainees at that workshop had psychiatric histories, and Caroline asked them whether this leap to the prospect of a murderer in their midst could have less to do with its probability than with a fear they'd absorbed. "Are these types of thoughts rooted in oppression, in the idea that folks with psychiatric conditions are a danger? It's easy to internalize that worldview. What is a person really saying when he says, 'I want to fucking kill my parents'? I've had groups respond by saying, 'It seems like you're in a really painful place. I've had those feelings. I've felt like that when I was hurting.' And then the person gets to name their feelings and be heard."

After the training, Caroline and I talked more about that moment. "It's not as recognized as other systems of oppression," she said, "but there is a deep prejudice at work. People have been in need of support, and what they've received is risk management. And it can be hard to shift people's ideas, even when, or especially when, they know the pain of being presumed to be unhinged, to be a threat, the pain of having the system in control over them. They will have a tendency to see others the way the system has seen them, to do what's been done to them. Think about how pervasive the notion of unhinged and dangerous is. Think about Halloween. Halloween tropes are full of people who've been diagnosed and hospitalized, people who are monsters. We need to call these things out for what they are. But for now, in a three-day workshop, it can be hard to get people to slough off those prejudices."

In the conference room, Caroline clicked on another slide: "We honor that suicidal thoughts are valid responses in people's lives." And another: "Alternatives to Suicide sees suicide as solution rather than as the problem." To the person contemplating killing himself, the end would be release. The thought of death was a balm. The problem wasn't suicide; the problem was whatever suicide was going to solve. The goal of the groups, she taught, was to hear all that felt insoluble. Gradually it might be right to ask the person if there was anyone, or any animal, or anything that had helped in the past. But this type of question was to be the extent of intervention. The mission, as another slide declared, was "to stay present," and not—with a red X through the phrase—"To prevent them from doing that."

She asked a trainee to read aloud from the screen: "The Guest House," by Jalal ad-Din Mohammad Rumi, the thirteenth-century Islamic scholar and Sufi poet:

This human being is a guest house.
Every morning a new arrival.
A joy, a sadness, a meanness,
Some momentary awareness comes
As an unexpected visitor.
Welcome and entertain them all.
Even if they're a crowd of sorrows
Who violently sweep your house
Empty of all furniture,
Still, treat each guest honorably.
He may be clearing you out
For some new delight.
The dark thought, the shame, the malice,
Meet them at the door laughing,
And invite them in.
Be grateful for whoever comes
Because each has been sent
As a guide from beyond.

AFTER THAT WORKSHOP, the pandemic closed back in. Alone in her office in the attenuating downtown, the disappearing province of the fragile, or walled off in her ground-floor apartment a mile away, Caroline taught trainings by Zoom and conducted HVN and suicide groups on her screen. She'd once told me that leading the groups was, for her, "a spiritual practice": the disheveled rooms were sacred spaces; the communal listening had the resonance of prayer. But it was hard to feel the sanctified in the flatness of a screen or the transcendent in the interaction of thumbnail images.

Yet she sat before a computer beyond the point of exhaustion. She had to make sure that despite the pandemic, seeds continued to be planted and seedlings didn't shrivel. And she had to tend to the flourishing demand that packed virtual rooms and engendered more meetings, because the pandemic intensified isolation and because the online gatherings were easy for anyone, anywhere, to reach.

At all hours, too, there were talks with lone people, by phone, by Discord. There were long emails back and forth. There was a grandmother in Kansas who needed to know that her grandson's voices would not destroy his life. There was a woman whose sister, a graduate student in primatology, wanted to be let go from the hospital but was confused about her rights. There was a young man whose house was under surveillance and who wound up speaking with her about the sources of his shame. There was a woman whose voice was commanding that she cut off her hand, that otherwise the voice would harm her child. Caroline steered her to thinking about what else the voice might be straining to communicate, might be suggesting, in a backward way, beneath its horrifying terms, might be struggling to teach about being a mother. There was a woman who, late one night, as Caroline brought their talk about her son to an end, begged, "When can I call you again? When? When, when, when, when?"

Haunted by the conversations and circumscribed by the pandemic, it wasn't easy for Caroline to fall asleep, when finally she silenced her phone and closed her computer and turned away from whomever was trying to make contact. The only medication she used was for this purpose: trazodone, which in higher doses was considered an antidepressant, but at the low dose she swallowed was prescribed to treat insomnia. It was a concession. She knew that without rest, she couldn't navigate all she heard. Yet now the drug wasn't enough. Hours passed; she remained awake. The sound of birds, the wings of large birds, a flock of them, beat close and loud, their feathers almost felt. The birds encircled and ensnared her.

Still, she woke on little sleep and began again, training, leading, listening, placing a yellow pin on the map for every new connection she made. There was a man, fifteen hundred miles away, who had Tourette's syndrome, who felt demons acting upon his body, and who needed help in finding the beneficent forces in his life. She guided him, and between his call and the next, she heard a beneficent voice of her own. It had appeared in the past several years, as if it had been waiting for her in Holyoke. This voice sometimes sang to the others, the others whose fierce protection, of her and themselves, and whose hostility and aggression, toward her and the world, and whose sheer anarchy arose, she had come to understand, from a distilled fear, an immense vulnerability and fear embedded within, embedded initially she couldn't say exactly how, but embedded permanently, elemental inside her.

"There are so many of us on the road," they said, and demanded, frightened, "Where are we going? Where are we going?" and "Who is going to feed us?" But the singing voice, a woman's voice, reassured them. "She doesn't sing in any genre that I recognize," Caroline said. "It's slow. It's old-time music." Attempting to demonstrate, Caroline hummed a single high, quavering note, throat worn with fatigue. "She's singing that they'll be able to put their roots down. She calls them 'my children.' She's telling them that they're not going to be on the road too much longer."

TWELVE

Dav020id asked if I knew anyone who had recovered from such anguish. His romantic weekend with Amanda had begun auspiciously. They had set out on Saturday for a hotel she had chosen in a town four hours from home, with the idea of hiking that afternoon and again the next morning, before returning to be with Gillian on Sunday evening. But halfway there, traffic slowed them, and then the sky darkened, and when they stopped at a diner for lunch, the air felt more like early winter than mid-fall. They realized that, given the hour, an outing that afternoon would have to be quick and might not be possible at all, and suddenly the long drive started to seem foolish: nine or ten hours in the car, there and back, for the sake of a few hours on mountain trails.

It was Amanda who raised this, but it was David who seized on it, as if any reason for paralysis was too seductive to resist. The failure of psilocybin and TMS, the futility of his own will, the persistence of his searing skin, the exhaustion of four-hour night after four-hour night—the allure of paralysis was compounded by the appeal of home, a place to hide, to cower. He was like a man who has been tortured and is driven by the need to conceal himself, in order to evade not only more pain but the abject shame of being unable to escape, incapable of controlling anything. He pointed out that if they turned around and stayed at a hotel they'd just passed, they could cocoon themselves and watch a movie, and that they could probably find a hike nearby for tomorrow morning. He was already thinking about the relief of reaching home early.

They left the diner and merged back onto the highway. They headed back in the direction of the city, and soon exited, crossed a bridge, and entered a village of sizeable, prefabricated houses, a surprisingly upscale shopping mall here in the middle of nowhere, and the hotel. None of it made any sense. Why plunk this new development of sprawling homes on streets ending in cul-de-sacs, chain clothing shops, chain restaurants, and a high-end grocery store in this spot instead of twenty or forty miles to the east or west? It felt like an arbitrary scar on a landscape of rugged cliffs and a churning river. And within minutes of their checking into the hotel and stepping into their pleasant, generic, and, they were assured, sanitized suite, he knew that their being here made no sense, either. What had seemed, a short while ago, a practical and cozy reversal of their original weekend plan now seemed like nothing but a reversal.

He knew he was fully responsible. Amanda had merely remarked on the crazy amount of driving they would be doing; he had been the one to turn them around. He could have said that they should do the crazy thing. He could have expressed a desire to be on those distant trails, never mind the weather and never mind if they had to cut a mile or two off those hikes. It was what she needed from him, she'd made clear: for him to throw himself into things, into *them*. But once again he'd let the self-absorption of his despondent state take over. It was stunning how his selfishness and loss of self were inseparably paired.

In their suite, they had a Zoom chat with another couple and, in front of their friends, began to bicker. The subject was COVID and whether they were as careful as they believed, whether they were honest with themselves about their lapses. When both couples clicked "leave," and David and Amanda were alone, their squabbling became a full-blown fight over their pandemic hypocrisies. It ended abruptly. "I just can't do this," she said. "I just can't do this. I need to separate myself from you." She slept in the suite's living room.

When David asked if I, in the course of working on this book, had talked with anyone who had recovered from anguish like his own, I told him yes. He was the only one to ever ask this of me. Others who were mired, whose battles were ongoing, never did. It was as if his mind—the mind, no matter his obliterating self-doubt, of a gifted litigator, a deft logician—was especially ill-equipped to endure the psyche's infinite uncertainties, all the answers it withheld.

Carl was one of the people I was thinking about when I said yes. His suffering wasn't bound up with medication withdrawal; in that way, his story wasn't relevant. But he had surely been in anguish. When he was thirty, he'd tried to hang himself from a tree in a park in Hartford, Connecticut, after leaving a note, which concluded, "The thought of the grief I may soon cause is difficult to bear, but that's easy to avoid: don't murder myself. That's the obvious choice. The correct choice. The far kinder choice. I wish I wanted to bring myself to make that choice."

Carl had been a psychiatric patient since the age of seventeen, diagnosed alternately with depression and bipolar disorder, given antidepressants and mood stabilizers, and prey to, he told me, onslaughts of mortification that were, he knew, thoroughly out of proportion to the mistakes or minor failures that incited them, but that were impossible for him to quell. He couldn't explain why his shame loomed so large. He couldn't trace it back to anything specific in his childhood other than his father's intimidating anger when he struggled, in third or fourth grade, with his multiplication tables, which seemed to him a flimsy explanation, and which plainly hadn't harmed him in the area of math. He was soon solving, in his head, complex algebraic problems and later doing the same with calculus. He assumed there must have been other incidents with his father that he couldn't recall, and he knew that his younger sister had graduated from high school early so that she could leave home, because her relationship with their father was volatile and came, once or twice, to the brink of violence, but the fact was that his memories were mostly of two loving parents.

The only other explanation, less for his periods of unmanageable shame than for the alienation and hopelessness that came with them, was his mother's personality. "She's outgoing, happy, bouncy," he said. "She never told me, 'Just snap out of it,' but there was always the implied question: 'How can you be so depressed when things are so fine?' It's hard to describe depression to someone who's never experienced it. Now I would say it's like you're in a boat on the ocean with no land in sight, and the ocean is made of pain. You don't want to dip your oar in the water, because you'll splash yourself and be in even more agony than you are already. So you sit and do nothing. And then the boat starts to leak."

At thirty, he set a date and scouted the park in his Hartford neighborhood for the right tree, one he could climb and one with an ideal branch. He chose a late-night hour, to reduce the odds that anyone would see and try to stop him. He left his note, indicting himself for failing to resist what was "easy to avoid" and for all he was inflicting on his parents and sister, though otherwise he took care to be considerate, "to minimize the collateral damage," he said, by calling the police just before he leapt, so that no one else would discover his dangling body. He was unconscious when they cut him down.

Six years later, he had a computer engineering job he was proud of and a fiancée he adored. He was not opposed to psychotropics—he figured they'd helped him temporarily after his attempted hanging and at various other times of terrible distress, times when "showering and then just getting out of bed went out the window"—and he'd always been fairly lucky with side effects. But he wanted to learn to live without the latest cocktail he'd been prescribed, and after a slow titration, he'd been off medication for eighteen months. He hoped to be but didn't count on being off psychotropics forever, and week by week, day by day, what seemed to buoy him, beyond his work and his fiancée, was, in his words, "a sort of pantheistic existentialist absurdism," along with a devotion to birds. He talked about his pantheistic absurdism with an intimation of laughter in his voice but also with a tentative faith. Behind his quirky spirituality was a recent proposition in mathematics and physics, the Free Will Theorem. "If any part of the universe has free will, then the universe as a whole has free will," he said. From this premise, he edged closer to a religious view: "And since free will seems to depend on consciousness, I believe that if I have consciousness, then the universe must as well."

He added, "Nonsense is a source of salvation," yet he said this not with a note of nihilism but rather with amusement at the tenuousness of his reasoning about the cosmos. Then he spoke of meeting his fiancée on a dating site where he'd featured birds in his profile; of their first date in the marsh of a nature preserve, surrounded by loons and geese; of his relentless efforts as a child to pet the pigeons in a park near his family's home; of how difficult it was to describe his allegiance to birds, except to say that he felt "love and compassion and joy" when he was with them; of his wish to become an avian vet until he realized how much putting sick birds to sleep would wound him; of the parakeets—one lemon-colored and the other white with dark flecks—who lived in a room behind his home office, always visible through French doors, and who were capable of rescuing him from depression; of the privacy of wood ducks and the conversations of the tufted titmouse; and of crows.

He showed me a video of crows. The crows liked a type of nut with a stubbornly hard shell. Their beaks were useless. But they had learned to perch themselves on a wire above a cross-walk on a street with heavy traffic. They dropped nuts onto the crosswalk, waited for car tires to do the crushing, flew down to the street, waited again for the pedestrian light to turn green, and hopped out onto the crosswalk to eat the meat of the nuts in safety. Here was more comfort than his tenuous reasoning could provide. Here, in the brilliance of the crows, was a hint of the conscious-ness of the universe.

Long after midnight, on the night of Election Day, David went to bed with the outcome of the presidential race unresolved, a lack of clarity that his mind turned into vivid scenarios of cops and military tanks confronting anti-Trump protesters in the streets and of his inability to protect them, his failure to so much as file a petition, his incapacity simply to gather witness accounts of First Amendment rights scorned and protesters beaten to the pavement and put behind bars.

When he woke, he was assaulted by the worst nerve-scorching he'd been subjected to since the symptom first flared across his skin, with the burning now encompassing not only his body, from his feet to his scalp to his tongue, but enflaming even his teeth. He wasn't sure whether teeth had nerves. He thought he could be go-ing completely mad. The media still weren't calling the election. Too many votes were uncounted; too many states were too close; too many permutations were mathematically possible; and while he was able to hear and comprehend that Trump's victory was im-plausible, this wasn't enough to calm him, because his nightmare imaginings erupted, too, from a Trump defeat that was narrow enough for Trump to viably contest, bringing out the marchers and police riot squads and the National Guard and Army Reserves, and pointing a spotlight on David's impotence. He never lost track of his drug withdrawal as the cause of his nerve pain, but it was as if Trump had access to his skin, to his bicuspids and molars.

Reprieve came by week's end, a retreat of the searing to regular levels and somewhat lesser territories, a confidence that the country had "spared itself," he said, "four more years of this lunatic." State by state, the margins looked solid enough that Trump would have no case in the courts, no means to drag out the process. In the streets, there was music and dancing instead of marching. It seemed that the transition would be peaceful.

Reprieve came in another form as well. Amanda didn't leave. And three weeks after the election, on another weekend away, they had sex for the first time in well over half a year. "Amanda booked a cabin for a night," he said. "The stated agenda was to reconnect, and it started about as awkwardly and awfully as anyone could conceive of." There was stiltedness followed by a squabble. There was her refusal of a massage followed by their sleeping in separate beds. But in the morning, in the last hour before checkout time, they succeeded. "It was a real relief. To kiss, to be physical with her, to fuck, it felt really good. I was so grateful. I brought her to orgasm, and that felt good. I had an orgasm, and that felt good. It was a pretty big deal."

There was, in those weeks, yet one more reprieve. At work, the organization held long meetings to consider cases and overarching strategy for the coming year. David conducted the discussions, and afterward he received, from his boss, compliments on his leadership. Then he told someone on the staff, a tech specialist he'd bonded with, a bit about his fear of lost competence as a litigator, lost passion as an advocate. The IT guy replied that he'd picked up just the opposite from David's colleagues. "One thing I can tell you," he said, "is you have everyone fooled."

Friends told him that they saw and heard a change, that his voice was lighter, his speech more swift. One of his internet support group allies, with whom he'd spoken by phone, said, "You've never sounded so good." One evening in the kitchen, Amanda cued up a two-part song of "Age of Aquarius" and "Let the Sunshine In," a favorite of her parents, and when she and Gillian danced to the 1960s anthem, he joined them.

"*Do* I seem better to you?" he asked me. "I know you're a writer, and this isn't your role. But can you tell me? Because better isn't necessarily how I feel. Because sometimes I feel like death. Does my voice sound lighter? Does my brain seem quicker? Or does it seem as impaired to you as it does to me? I know you're a journalist, but can you tell me what you perceive?"

MY BROTHER MET the love of his life while he was living at the homeless shelter. He called me at around the same time. I don't think he said anything about Pam. He didn't speak about all he'd just been through or about where he was living. He was calling to apologize. He said that he wasn't going to be at my wedding, where he was supposed to be my best man. I'd asked him to take this role many months earlier, while he was still in D.C., and I was bewildered by his words now, but I was less upset than relieved.

I didn't need to know that he was sleeping in a shelter and eating in a church basement and washing his shirt in the shower in order to worry about how odd he might seem to my friends. Among the guests my fiancée and I had invited, corporate law was the dominant profession. My bride-to-be was about to start work as a federal prosecutor in Manhattan, a plum position and rite of passage to a powerful legal career. Our social world was conventional, low on idiosyncrasy, barren of strangeness; it would take years for me to pull myself away from the comfort and concealment that convention offered. But my brother was someone who had choreographed and performed a dance on ferry decks, who had believed he could save our grandfather from Alzheimer's, who had been in a mental hospital. Who knew what he would look like at the wedding—what he might wear; whether he might arrive respectably kempt or in disarray—or what he might say to our friends, or whether he might decide to give a trial run to some bizarre dance?

I tried to convince him to change his mind, but I didn't try hard, and in any case, he wasn't persuadable. He said that he didn't want to be in any proximity to our parents, that their effect on him was destructive, that it was malign, that he couldn't count on being around them at my wedding and emerging intact. Such seeming hyperbole made me feel all the more that it would be easier not to have him there, that he would be prone to spouting this sort of talk at any moment.

Yet he told me later—and I have no reason to doubt him—that the kind psychiatrist at the run-down state facility had listened to his story and suggested that he *should* stay away from our parents for a time. In other words, this psychiatrist, who did try, gently and unsuccessfully, to get Bob to agree to go back on lithium, nevertheless believed that Bob's perspective on our parents was rational enough, that his view of our parents as a malign force in his life might not be senseless at all.

Bob met Pam in the offices of Harvard's dance program. He had taken the forty-five-minute bus ride from the shelter to Cambridge and managed to get himself an audition to accompany the university's dance classes. The audition was held in a nineteenth-century ballroom that served as a dance studio. There were candelabra sconces and gargantuan chandeliers. There were fireplaces with elaborate mantels and a gold-framed portrait of an eminence from Harvard's past. Floor-to-ceiling windows were bordered by dark, intricately carved wood. Outside, a storm was on the verge, a gray-green light pressing itself to the glass and permeating the room. At the professor's command, the students set themselves, bare feet on the ballroom's amber-tinted floorboards. She counted and clapped out a beat, and Bob's task, at the piano, was to inspire the dancers while never betraying the teacher's cadence. As for genres, melodies, styles, he could play whatever he wished, any piece, any improvisation, any message he was moved to send to the dancers' bodies.

The light shifted lower; it was nearly night in the room; thunder rumbled; the downpour began; it seemed that the windows would be too weak for the water. He played into this merging of outside and in, played a jazz piece he had written. Dissonant chords chiseled fissures in the melody. His fingers were in unremitting attack on the keys. It might have been only the weather's tumult, the way the lightning suffused the studio, but the dancers seemed swept away, and they remained transported as he switched from jazz to classical, from improv to formality, though with little slackening of assault, no lessening of alliance between music and storm.

He was hired. Pam was the administrative assistant to the program; she gave him his employment forms. She also co-taught a modern dance course and performed with a small but notable New England company and would be hired, within two years, by a Canadian company that would fill the Joyce Theater, one of the premier venues for dance in New York City. To a packed house, she would dance the lead in the company's reimagining of *The Rite of Spring*, the experimentally jagged and erotically ravaging ballet that had exploded the worlds of dance and music on the eve of World War I, and when the performance was over, night after night, she would receive a standing ovation.

This is to say that she, a tall, reserved woman with a pageboy haircut and wearing a long skirt, both was—and wasn't—a program clerk who gave him his paperwork to fill out. She, who chatted with him whenever he showed up in her office, and who heard much of his story over beers on their first date, and with whom he remembered sharing "a prolonged clasping of hands" afterward as they said goodbye at a T station, and who wasn't particularly concerned about where he was living nor about his stints on psych wards, was an artist of rare talent and drive and no reserve whatsoever when performing, a woman of taut muscles and an analytic mind and angular, beautiful features who was already making her way as a dancer and was about to own one of dancing's major stages.

On their third date, after they ordered their hamburgers, he took out a spiral notebook and turned to a page where he'd printed in pencil and read to her one of Shakespeare's sonnets.

Let me not to the marriage of true minds
Admit impediments. Love is . . .
An ever-fixed mark
That looks on tempests and is never shaken;
It is the star to every wand'ring bark . . .

And when she joined the Canadian company, she took him to live in Montreal. She had roots in Canada; her mother was Canadian and moved to Montreal to be with her. Pam and her mother attended an Anglican church, where Bob found work as the director of the choir and as a singer. He sang the bass solos in Handel's *Messiah* and performed Bach's aria *Quia fecit*—"Because He who is mighty has done great things for me"—and sang the raucous spiritual that told the story of Joshua's canonlike, miraculous trumpet: "Joshua fought the battle of Jericho, and the walls came tumbling down!" His musical arrangements, his piano, his voice swelled within the hundred-and-fifty-year-old stone sanctuary. That was on Sundays. On Saturdays, he sang at synagogues, one of them across the street from the church, sang a solo whose words had been written two and a half millennia ago, in the desert of what is now southern Iraq, a prophet's attempt to reproduce the reverence of angels.

My brother and Pam were married in Montreal's central courthouse, with a small reception afterward at their apartment, during one of the harshest blizzards ever to hit the states from Virginia to New York. Montreal was beyond its periphery, but by the end of the three-day storm, Maryland, where our parents lived, was covered in three feet of snow and struck by winds that created drifts twenty feet deep. Wildly rising tides flooded New York City's subway tunnels and shut down the system. Airports were closed; train tracks were buried.

Yet as the weather forecast grew ominous, and then as the storm got started, and as it seemed there might be no way for our father or me to reach Montreal, our father told Bob that he should not delay. A courthouse wedding with a few friends cooking food for the reception could easily have been postponed for a week, but our father insisted that there should be no postponement.

Our mother was, at that time, under hospice care. She was restricted to bed and reduced to painkillers. She had been diagnosed with ovarian cancer three years earlier, cancer that had already, before her diagnosis, metastasized to her stomach and liver. She had endured chemotherapy and surgeries that lent no more than fleeting optimism. My brother and our parents were in contact again, barely. Her paramount wish, when she finally admitted to herself that more treatment would be futile, was to stay mobile enough for long enough to be at the wedding. And when her strength plummeted, when intravenous feeding no longer made sense, when her paramount desire became undeniably impossible, she made it clear, with all the emphasis her shrunken body and desiccated throat allowed, that our father was to be there for the ceremony, no matter what condition she was in, no matter if she was scarcely breathing and it seemed that she might die while he was gone.

He gave her his promise. Devastated as he was to be losing her after forty years of marriage, destroyed as he was by the prospect of not holding her as she died, of not resting his forehead or pressing his nose against her cheek as her breath thinned to a wisp and went still, he agreed with her. One of them needed to witness the marriage, he told me, to be present and lend solidity. To them, the wedding of their son to Pam was too much good fortune to be trusted. How had their diseased and unmedicated child found a wife who seemed so levelheaded, so directed, so gifted, so dependable? It seemed she might be his safeguard. It felt, to them, too farfetched to be true, and it was essential that one of them be there, to make it more official, more real, somehow less precarious. With a father's and mother's love, they yearned to diminish the acute fragility they perceived—and lived with—in their son's life.

When the blizzard got under way, our mother was still fully conscious. She was part of the decision to tell Bob not to postpone the ceremony. For her, the worst outcome, as she died, the outcome unbearable to imagine, would be to know that there hadn't been a wedding, that it had been put off, that there were no guarantees, that it might drift, could be delayed indefinitely, that there could be some break, some fight, some erratic behavior on his part, or something worse than erratic, something to fill Pam with foreboding, and that he would lose her and be lost.

As the snow started to thicken in the air and accumulate on the ground, and as she rested, eyes closed, still breathing, she was perhaps able to keep vague track of our father's frantic efforts to find an airport that was still open and that he had half a chance of reaching and that had some route, never mind how circuitous, to Montreal that day or evening, so he could be at the wedding in the morning. Perhaps she was aware that I was doing the same in

New York. I could find nothing. The airports around New York had surrendered to the wind and snow.

But our father stepped out into the blizzard and into his car and somehow stayed just ahead of the heaviest snowfall and navigated to an airport that was functioning. In the morning, he called our mother to say that it had happened, that it was true, that it was done. A family friend held the phone to her ear. The friend maintained, to us, that she heard his voice and surely understood. Our mother lost consciousness shortly after that, and soon stopped breathing, without her love of four decades, having sent him off on a mission that was more urgent, more imperative to her, as Bob's mother, than dying with her husband's hand gripping hers, his skin against her skin, flesh against her dry flesh.

ON JANUARY 6, rioters stampeded the Capitol. David, working at his computer at home, read fragments of descriptions of what was happening, but evaded full awareness until Amanda called out to him for the second or third time, telling him he needed to join her in front of the television. Then he saw Trump's devotees thronging the steps and scaling the outside walls like frogmen. He saw them pummeling cops and bearing the Confederate flag. He saw them smashing windows and ramming doors and forcing their way into the Senate chamber as Congress was supposed to be going through the formality of affirming the results of the presidential election.

That day had a different effect on him than the days leading up to and just after the November vote. Two months ago, he'd been overtaken by fear that he would be incapable of doing what he was—or had once been—called to do, that he would prove useless in defending the rights of peaceful protesters in the aftermath of a close and contested election, that he would be exposed as incompetent and impotent and that his career as a champion of civil liberties would come to a humiliating end. January 6 demanded nothing of him. He wasn't about to argue for the rights of the rioters. But the images frightened him in another way, a feeling tied to what had flooded him when Trump had first taken office. "I'd thought," he said, "that there were some basic truths that could be counted on, that people like Trump didn't get elected president, that there was a certain stability. But in the place I'm in, my own personal country of no resilience, watching the footage of the riot made me totally uncomfortable. If I was healed, the world probably wouldn't seem quite so scary. I know the two shouldn't be linked, that part of this is the narcissism of my condition. But watching the riots was like when I was a boy, and we had this banging radiator in our apartment, and at night in my room I was petrified that people were tunneling up to harm us."

He wondered, after so much wrestling, whether he should give in, go back on medication, sacrifice all the time and suffering he'd put into letting his brain regenerate itself. He went to a sleep specialist, who recommended trazodone for his insomnia, the drug that was Caroline's only acquiescence. He went to a psychiatrist, who suggested that it was time for him to return to medicating his anxiety and depression, and who coaxed him to try a type of antidepressant that had been sidelined decades ago, because it interacted chemically with a list of food and drink, from cheese to tofu to draft beer, in ways that could be fatal.

David resisted the antidepressant; he wasn't ready to concede defeat in what had been his primary quest, to find out who he was without a pill he might never have needed. And there had been signs of something more authentic emerging. "Close friends of Amanda's," he said, "have told her that I seem more present than in all the years they've known me, that we're having more interesting, real conversations." He counted: seven people had commented, in some form, about the person crawling into being. He put off any experiment with another antidepressant and tried the trazodone for the sake of sleep, figuring that the recovery of his brain was impossible on his current hours of rest. The drug did nothing. The sleep specialist doubled his dose, with minimal, seductive, then disappearing effect. The doctor said that an even higher dose was warranted. David refused and was left with another medication to wean himself from, snipping milligrams from the tabs with his pill splitter, unable to mute his worry that his sleep would rebound for the worse, which, inevitably, it did.

He kept up his swimming, his ukulele lessons. "I'm going to play you something," he told me, in his little outbuilding. "I've made progress. My teacher would probably say not a lot of progress for the length of time I've been taking lessons. And we'd agree I'm not much for natural talent. But I've gotten better. I'll play you this. You'll know what this is."

He struck a chord, strummed a deliberate rhythm, fingered another chord, climbed to a higher plane. He repeated the sequence. His cadence was steady. Not singing, he played—strumming down-up-down-up-down-up—the entirety of a song that, whether for his shortcomings or my own, I could not identify nor even hear as a melody. I hedged when he asked if I knew what it was, and quickly he saved us. "That was 'Somewhere over the Rainbow,'" he said.

In April, at the trial of Derek Chauvin, the policeman who'd killed George Floyd, the prosecution rested. The defense was likely to be brief; jury deliberations and a verdict were near. Eleven miles from the Minneapolis courthouse, a cop shot and killed another unarmed black man; protesters amassed outside police headquarters; journalists covering the rallies were punched by cops, splayed on the pavement, detained. In Los Angeles, the closing of a homeless encampment stirred resistance; reporters were cuffed. "This is a reprise," David said, thinking back to the period before the presidential election, when he'd dreaded having to file emergency petitions to protect marchers' First Amendment rights. Now, more and more it seemed, journalists were under attack. And if the Chauvin jury came in for manslaughter but not murder, or for acquittal on all counts, or if the jury was hung, and if journalists were arrested as they tried to film and write about the uprisings in his region . . .

"I am frozen like a deer in headlights. I have no idea what we would do," he said, though he knew precisely what he would do: direct his team to collect testimonials from reporters; garner expert opinions on the unconstitutionality of the arrests; write the pleadings. And he was already composing a letter to police departments admonishing that if they declared unlawful assemblies, they were obligated to let journalists stay to witness the events. But despite his plans and preparations, as the defense called its witnesses, he said, "This isn't last fall. Bad as that was, then I had more wherewithal. Now I'm so terrified, I'm an ostrich."

The defense rested its case. With closing arguments and jury deliberations days away, he fixated on what he should do, immediately: call his boss, confess that he could not cope, put himself on disability, and let someone else take over. "If shit hits the fan and journalists get abused, I will feel guilty if I'm not there, but I will feel guiltier if I'm there and in everyone's way. An hour ago, I was out in the studio shed, on the floor, weeping, saying, 'Please please please help me, please please please rescue me.' I can't keep pretending. Who am I? A hundred times a day I ask myself how much of this is truly me and how much is the withdrawal. A hundred times a day I date the insomnia and the burning to going off the meds, and I tell myself the depression was never, ever this severe. I tell myself that even in 2016, 2017, I would come home from work and see Amanda and Gillian in the kitchen and think, *No, everything isn't perfect, but this is my wife and this is my daughter*; I felt a contentment; I knew that my work meant something; I knew our kitchen as a place of love; I may have been a troubled person, but I wasn't who I am now; and I still tell myself this isn't truly me;

and I believe it; and I tell myself it's a lie; and I tell myself it doesn't matter, because it's like we're made of glass, and once you break that glass, shattered is what you become."

He lectured himself not to abandon his work as the jury deliberated. It would be difficult for someone else to step in days or hours before the verdict. But as soon as Chauvin was convicted of murder, David went on disability.

"IT DIDN'T GO away," my brother said. "When I was playing for dance classes, when I was playing or singing in church or synagogue, I would think, *I better play well, I better sing well, or they're going to find out I'm crazy.* I would be *at the piano*, thinking it. All these people with initials after their names have given you this label, have told you you have this, and your mother has told you you'll kill yourself if you don't treat it, and your father has told you the same, imprinting it deeper, with his eyes, imprinting it every time he's looked at you."

About the way our father continued to see him, Bob wasn't imagining. Our father said to me, when my brother was well into his thirties, "We need to treat him—" I can't recall the word he chose. Carefully? Tenderly? His tone is what remains with me, his tone that expressed a vision of Bob's psyche as a house of cards, susceptible to collapsing in a mere movement of air.

Beyond the music of worship, beyond singing the slow, ascending, incantatory notes of Fauré's "Libera me"—"Deliver me, O Lord"—in a way that seemed to seep into the church from afar before enveloping the pews, beyond Sunday and Saturday mornings, he played for dance classes at one of Montreal's universities, studied dance, performed as a dancer here and there, joined a new improv company with Pam. But all the while there was another pull, an alternate gravity, again the presence that had materialized, immaterial, on the Olympic Peninsula, verbal now, saying to him silently, definitively, *All you have to do is say yes.*

He heard it during a delayed honeymoon, in the Old City of Jerusalem, in a twelfth-century church where one afternoon he sang to himself in the empty sanctuary, and where the nuns in charge locked him inside when it was time for their siesta. Had they failed to notice him? It didn't seem likely. Had they chosen to leave him alone in their house of prayer? Pam was elsewhere, touring other sites. He sang openly in the church, famed for its acoustics, its pale stone, in hues of pink and beige, able to turn a lone voice into a Gregorian choir. Etched beneath an archway was an Arabic inscription. For centuries, Muslims had prayed here five times a day. Outside were the remnants of a Roman temple and the rubble from a statue of the Greek god of healing and the walls of a pool where, it was said, Jesus had cured an invalid, empowering him to walk.

You've been walking toward this your whole life.

Had he? He recalled, from childhood, sitting at our grandparents' table at Passover and listening as our grandfather read in Hebrew, read incomprehensibly, interminably, it seemed to me. But Bob was transfixed by the rhythms, spellbound by something beneath the words, struck by something in the way the gold-tasseled cloth lay covering the unleavened bread, and by something in Clara's lighting of the candles and praying with her hand delicately covering her eyes.

In the north of Israel, on the Mount of Beatitudes, he sat meditating, breathing, surrounded by date and palm and cypress trees and overlooking the Sea of Galilee. He believed he caught a glimpse of the trees and water as Jesus had seen them, a fractional, infinitesimal share of that long-receded, long-vanished biblical moment.

All you have to do.

The presence accompanied him on a solo backpacking trip in the mountains of Quebec, and on his many walks up Mount Royal in Montreal, up to a grove of pines that seemed, though it was in the middle of a city park, to belong to him. No one else was ever there. No one so much as passed by. He sat on a cushion of pine needles and practiced the Hindu breath patterns of pranayama. He did this throughout the year, sitting on the snow, his body still, the snow and extreme cold creating a limitless stillness in the world, as he inhaled and exhaled with no insinuation of discomfort, despite the temperature, with, instead, "a sense of the interwovenness of the universe, a feeling of connection to all that was and all that is."

That was how he recounted it to me. There had been no discomfort in the cold, but there was unease in his voice as he said "all that was and all that is," an unease that, at times, took hold of his inflections when we talked, the unease of someone confiding an experience that may not be understood or accepted.

At one of the synagogues where he sang, he met with the rabbi about studying to be a cantor. The rabbi was all encouragement, all praise. At church, he delivered a lay sermon. It was a parable he had written about a man's private battle between his rational objections to one of Christianity's core tales and the irrational consolation he finds in the story. Afterward, clergy asked if he had ever considered seminary. He preached again on another Sunday. They said that if he would enroll, his tuition would be taken care of.

The ending of *Franny and Zooey* returned to him. Its lesson was newly distilled by him. It was phrased in its humblest form: "You'll do what you do, what you can, what's given to you, and it will be a little bit helpful to people."

The door is wide open.

He meditated on the roof of his and Pam's apartment building. He sat on a meditation bench six inches above the tar. Winter had passed; Montreal's late spring had melted the rooftop snow, though snowbanks still lined the streets. He channeled his breath to the chakra at the base of his spine, to the chakra near his heart, to the chakra at the front of his throat.

All you have to do is step through it.

The presence, without warning, was aggressive. There had always been a passivity; it had seemed willing to wait forever. But now, as he attempted to breathe, it was insistent, berating.

This is what you need to do.

He replied that he had his doubts.

This is the fulfillment of what you've been longing for.

He answered that it had never been clear to him. But he felt ambushed, attacked; the rhythms of his meditation had been replaced by inward trembling.

The path is there for you!

The presence, he realized, had already chosen *which* religious path, and now he lashed out, saying it was wrong, that he would not take that direction, not with our family's history, not with half of Clara's siblings and her mother exterminated. He could not. No matter how much he was drawn to Jesus's message, to his teaching of perpetual forgiveness, his embrace of the outcast, his elevation of the least, his laying of hands on the sick, his rejection of all dogma, his request of nothing more than faith, no ritual, no conformity, his healing of a divided world.

His body was in turmoil. His protests grew less and less articulate.

For you, it is the path of no resistance at all.

His internships during seminary, and his first job as a pastor, were in prisons. A hundred and ten miles north of Montreal, where winter temperatures plunged to negative twenty, he played guitar—one of a dozen instruments he'd taught himself over the years—and sang each week with sex offenders, the least of the least. At another facility, he arranged a room of fifty convicts in four-part harmony. He led a circle of inmates in meditation. He closed his eyes and guided them to pool their breath just below their root chakra. Increment by increment, he asked them to send their breath to the chakras above: slightly beneath their navels, below their ribs, along their brows, above the crowns of their heads.

For a year and a half, at the same hour every week, a convict came to him to prepare for his upcoming release. In a chaplaincy office that was nothing more than a cell, the convict confessed the gruesomeness of his murder. He needed Bob to know the kind of person he was dealing with. Then he asked for Bob's help in planning out precisely how to improve his chances of never committing another crime. Another inmate explained to Bob that he'd already accepted the Holy Spirit before he'd done terrible harm, and so, he reasoned, he'd assaulted the spirit within himself and was beyond forgiveness. Week after week, he argued this doctrine of permanence. Week after week, Bob offered the opposite argument.

In a cinder-block chapel with no pews, no symbols, no adornments, nothing to identify it as a place of worship, inmates gathered as Bob sat at a portable keyboard on a foldout stand. A convict sat beside him on guitar. The men taught him a country gospel tune, "The Old Rugged Cross"—"Emblem of suffering and shame"— and they all played and sang together.

When Bob and Pam decided to return to the States, he became the chaplain at a home for the aging. Researchers were just discovering that Alzheimer's patients who no longer recognize or remember their closest family still know the lyrics and melodies of songs that moved them when they were young. Bob started a choir in the home's memory-care unit. Light from an adjacent sunroom bathed the faces, sunken and so often vacant, but now, at the choir's hour, alert and animated and emitting sporadic laughter. Bob led them on the piano and cued them when prompting was needed. They sang "I Love Paris in the Springtime" and "All of Me."

My brother is, nowadays, the pastor of a church in a town outside New Haven. The wooden building, two hundred years old, is small and simple, with front doors painted a bright red and a white façade punctuated by narrow arched windows. There is a bell tower framed by understated columns and a widow's walk bounded by a graceful railing. The architecture of the sanctuary matches the exterior: a beauty of age rather than opulence or intricacy.

Because the town has no homeless shelter, my brother has unlocked the church doors on evenings throughout the winters. He has welcomed the homeless to spend their nights within the sanctuary's warmth. Another part of his ministry is a collaboration with nearby police departments. It's a kind of street theater and an adaptation of Isaiah's prophecy of a time when warfare will end: "they shall beat their swords into plowshares." From New Haven to Norwalk, Bob and an Episcopalian bishop raise awareness for gun buyback programs by standing on the sidewalk outside police stations on Saturdays with a forge on a foldout table and an anvil on a heavy block of wood. The forge is a cylinder of roaring gas fire, and the anvil is a massive, odd, ancient-looking object of black steel. Bob and the bishop, in blue jeans and loud orange tee shirts, heat and hammer pieces of handguns and semi-automatic rifles that have been turned in. They melt and pound and fashion the weapons into gardening tools. The trowels and clawed cultivators and hoes are laid out on display and will be given away later to school and community gardens. Bob sometimes takes a break from the forge to play one of the flutes they have made from the gun barrels.

The church congregation is a mix of New Englanders and first- and second-generation Nigerian immigrants, and on occasion, Bob turns over a Sunday service to the Nigerians. They sing Nigerian hymns and play the beats of home on a boom box. The women are in headwraps that flare into a fabric plumage of golds,

greens, pinks, purples. Their sashes and ankle-length wrap skirts
are resplendent. Their beads are thick and bracelets layered. The
men wear three-piece suits or traditional tunics. They beckon
the New Englanders, in windbreakers or fleeces, in stained white
sneakers or brown walking shoes, to follow them in a procession of
dancing—the movements taut among the Nigerians, timid among
the New Englanders—up the crimson carpet of the center aisle.

Before the pandemic and now again, Bob drives to a halfway
house in New Haven and steps into a communal room of beige
tables that looks like a sparsely populated school cafeteria. Ten or
twelve residents sit with expressionless faces; two or three rest their
heads on the tables and hibernate. Bob carries an acoustic guitar
in a case and a stack of songbooks he has compiled and bound in
faded blue covers. The covers are bent and abraded at the corners;
the books have been used many times before. "Hello, Manny," he
says to a man with a shaved head whose gray sweatshirt proclaims
"Cartel" and who says nothing in reply, keeping his back to my
brother as Bob unzips the guitar case. "Manny, can I give you a
songbook?"

Nothing.

As Bob moves around the room, this interaction is repeated,
with a woman who wears a large starfish pendant glittering with
glass diamonds, with a man too wide for his wheelchair, with
a slender man in an Oxford blue button-down, with a woman
whose red bangs shield her eyes. Bob greets those he has met on
previous visits, receives silence in response, and leaves a book on
the table in front of them.

"If it's okay, I'm just going to lay a songbook here," he says to
those he is seeing for the first time, "in case you decide to join us."

"Join us in *what*?" a man perks up from his stupor, his question
at once curious and hostile.

"Singing!" Bob says. "We're going to sing."

He announces that they'll start with "Lean on Me." He waits for anyone who might want to find the page. Manny and the woman with the starfish pendant, whose back has also been turned, pivot partway and grow louder as they sing along. One or two others join tentatively. But a woman in a hoodie is muttering to herself, then mouthing the words mockingly, then singing them in ridicule, before she strides across the room, declaring as she leaves, "I don't need no one to fucking lean on."

"You remember what I requested last time?" Manny asks.

"Yes, I do."

Manny, who appears to be Hispanic and in his twenties, seems an unlikely fan of the Clash, but for him, Bob has added "Rock the Casbah" to his repertoire, and the black, white, and brown residents—the seven or eight of them who have risen from their almost comatose states—murmur or come close to shouting out the exultant lyrics, with Manny leading.

Someone calls out "'La Bamba,'" and Bob launches into the opening run of notes. He relishes the line, playing it an extra time or two. A tall, bulky woman in high-heeled boots makes a loud entrance, glares, spins, exits, spins again in the doorway, and comes back into the room. She makes a face of seeming distaste and stands, eye to eye, squarely in front of Bob, who suddenly looks precariously balanced, standing with one foot up on a chair. He's switching fast between strumming chords and the irresistible run, and much of the room is singing outright. The woman clenches her fists and rolls her shoulders; she breaks into the tight steps of a dance, with mini kicks and gyrations. "One more! No, three more!" she cries out, when the song is done.

But food carts with dinner trays are being wheeled in. It's time for Bob to go. The voices of Manny and the starfish woman, Bob and the dancer, the man in the wheelchair and the man in Oxford blue and the woman with the veil of red bangs and a few of the others come together on "Amazing Grace," after which everyone returns to their initial indifference. Bob zips his guitar into its case, collects his songbooks, and leaves. When he says, "It was great singing with you," there is no answer. His goodbyes are met with silence.

Except that he is a man willing to stand outside police stations, making garden tools and flutes, and except that he conducts sing-alongs in a halfway house, you would not, today, notice anything unusual about my brother. In no way would you see a trace of someone labeled as severely—and permanently—mentally ill. You would see a broad-shouldered, solidly built man, a shade under six feet tall, sixty years old, with a shaved head and a face that could pass for forty-five. If you met him, you would chat the way people chat. He might say, "I'm the pastor of a church outside New Haven," and even if you are a person who finds a bit of craziness in that alone, in religious pursuits, it would never occur to you to suspect that a diagnosis of craziness might lurk in his past.

He and I have talked about the questions his life raises and what his story means. There are moments when he bundles it all up in a neat package. "I'm just lucky," he says, "that I'm crazy enough to have refused being crazy." Sometimes he touches on a brief period during seminary, when harsh doubts about the direction he'd chosen led him to wonder if he was indeed sick, to interrogate himself ruthlessly, and to contemplate going back on lithium.

Sometimes he mentions his regimen. Though he rejects any diagnosis of disease, he has long abided by a strict protocol for well-being. Daily he meditates and prays, does some form of arduous exercise, often biking up steep hills, and immerses himself in the practice of his instruments. On certain mornings, around dawn, he lets himself into a New Haven church near his and Pam's house. He plays the pipe organ while daylight rises to the stained-glass windows.

Our conversations can be expansive, extending into emails he sends the next morning with added thoughts and emendations. I told him recently about the parable of the turkey prince, and he wrote that he wished someone—our parents, a practitioner—had met him "under the table," met him where he was, been interested in meeting his mind, cared to listen in depth, instead of categorizing him, with great and swift authority, as someone doomed to the life-long terms of a dire illness. But reckoning with the infinite range of individuality, he pointed out, is not something that psychiatrists, or people, are genuinely asked to do or generally rewarded for doing.

Along these lines, he explained an idea for an upcoming sermon. He was just starting to shape the central metaphor. "Decades ago," he wrote, "a musical colleague told me that if you listen to chords closely you can hear the individual notes and overtones ringing out. I tried to do this with no success and decided to ignore his claim as either false or the result of a gift that he had but that I and most others do not. Yet I've found that it is entirely true. If you play a chord on the piano, and patiently listen to the sounds that are subtly produced, you can hear them all. They shimmer separately and in communication with each other. Knowing this phenomenon can transform the way you relate to music, yet the phenomenon is inaudible to most everyone unless they choose to open themselves to it and concentrate on it, unless they consciously train themselves to listen in this way."

Psychiatry—and especially biological psychiatry—may be inherently impaired when it comes to such listening. This is, partly, through no fault of its own. Science and medicine require systems of knowledge and treatment, systems of seeing and understanding, diagnosis and intervention, and systems aren't kind to individuality. They can't be. To commit oneself completely to individuality would be to renounce classifications and categories, and this would be to renounce science.

But partly, too, psychiatry is impaired by fears, terrors we all share, of difference and despair and danger, so that the profession's reflexive reaction to distress and to divergent realities, to life's agonies and its precipices, is to provide whatever medication is available, and to urge its long or permanent use, no matter how flawed the drugs, no matter how often futile, and no matter how potentially damaging—because to acknowledge medicine's predominant and persistent failure in the realm of the mind, to think hard before prescribing, to relinquish the edicts of protracted or lifelong medication, to surrender the illusion of control, would be intolerable not only to biological psychiatry but maybe to most of us, with our yearning for immediate solutions and the promise of safety.

And then, psychiatry cannot fully hear individuality so long as the profession clings to scientific authority. To listen, to truly listen, the profession would have to let go. It would have to embrace the idea of working *with* patients, of proceeding on footing that is more equal than not, even when *with* is elusive, even when it is inordinately difficult and when a stance of authority seems most warranted. The profession would have to embrace the idea of working in illuminating ignorance. It would have to acknowledge to itself that, as far as science is concerned, the mind is much, much farther away than the moon ever was, immeasurably, ineffably so. It would have to reiterate, at the start of every session, some variation on *We don't know*, and it would have to make clear to the primary care physicians who also prescribe psychotropics, and to all who inform public opinion and expectations, and to the public itself, what neuroscientists like Goff and Nestler have begun to pronounce: *We don't know*. Perhaps these three words should be psychiatry's mantra. Perhaps this might take us to a new beginning, marked by a new humility and enabling a new depth of listening and communication, for the profession and for ourselves. Perhaps by breathing and repeating *We don't know*, we might overcome some of our fears and understand one another more profoundly.

MY BROTHER AND our father are unhealed. For a time, before Parkinson's ravaged our father's mind, leaving him bewildered and frequently incoherent, when the illness had only crippled his body, my brother gave him singing lessons. This was a way to combat the weakening of his voice, its withering. Bob did this over the phone. Or he drove down from New Haven to our father's apartment in Manhattan and sat at the same piano they'd shared when Bob was a child, the piano on which our father had played, in his deliberate style, his habitual upbeat tunes, attempting to banish uncertainty and pain. With our father in his wheelchair beside the piano, Bob struck notes and asked our father to match them with his voice, coaxed them from his throat, implored our father to sing with him, a show tune, a jazz standard, slid his voice nimbly beneath or above our father's, singing gently in support.

But for my brother, the reverberations of the past are not gentle. Last year, when our father's mind had become only intermittently reachable, when his body had become all but paralyzed, when his face contorted in terror whenever anyone tried to move him from his wheelchair, Bob played a song for me, a song he had written, played it with a simple rhythm and little elaboration on the guitar, and with a short series of notes, plaintive and regretful, on the harmonica, and sang it in a voice that was inward, only faintly sonorous. It was less a performance than a private reaching out:

> I don't know if you know how to do an ending
> Or if you'll just fade away
> I don't know if I know how to do a sending
> If I'll have anything to say

Either way the arc of time is bending
We're surely coming to that day
I don't know if the bank of love is lending
This might be a good time to pray

It's hard to see you in your rolling chair
It's hard to have to help you down the stairs
It's hard to not hear you when you're talking
It's hard to see you looking scared

But really the hardest thing of all
Is I don't really want to call
Cause I know you know you nearly crushed my life
And I don't feel any regret from you at all

I don't know if you know how to do an ending
Or if you'll just fade away
I don't know if I know how to do a sending
If I'll have anything to say

Either way the arc of time is bending
We're surely coming to that day
I don't know if the bank of love is lending
This might be a good time to pray

Every week, my brother visits a psychiatric hospital. Broad-shouldered though he is, when he walks through the lobby, and as he waits for the elevator, and as he approaches the nurse-receptionist, in her booth behind plexiglass outside a locked ward, he looks somewhat frail.

It is a frailty borne of difference. He is doing something no one has asked him to do, something that is his own, that didn't exist in this place before he began. With his guitar in its case slung over his shoulder, he appears vaguely eccentric, a would-be musician in late middle age wandering around the facility.

Through the plexiglass, he identifies himself to the nurse. She is baffled.

"You don't remember me?"

She doesn't.

"Nurse Priscilla didn't tell you I was coming? I'm not on the schedule?"

"Un-uh. She's not here today."

Though he has been coming regularly, it is as if he is completely unknown, as if, between the changing shifts of the staff, the demands of their work, and the slight strangeness of his mission, he is erased after each visit.

But before he is turned away, a nurse happens to emerge through the heavy door into the vestibule, someone who remembers him. A call is made, a code is pressed. A male staffer, six-four or taller, in blue scrubs and tan boots with the laces untied, opens the ward door, which is marked in big letters, "AWOL PRECAUTION." My brother steps past him and down a corridor. The staff basically ignores him. He's left on his own to get the attention of the patients, all of them in their teens or twenties, in slippers and scrubs of a blue that's paler than the staff uniforms, with ID tags on their wrists. "Do you want to come sing?" he asks them, as they lie on their beds, or sit within side rooms, or shuffle along the hall. "Come on and sing."

"Do I want to do this?" a young man asks him, quizzical.

My brother's answer pulls the patient into a communal room of pastel-colored plastic chairs. A few more follow. Bob hands out his well-worn blue songbooks, and a teenaged girl flips through. "Very hospital-friendly," she says disdainfully. She heads away up the corridor. But the rest remain, and the number grows, and though they are mute, my brother's dexterity on the guitar holds their interest, until one of them can't resist and calls out the name of one of the ballads from the book. It is a new addition. It was a hit a few years ago.

"Do you remember this song, Dad?" a young woman with an ID bracelet asks the man in street clothes in the chair beside her.

He tells her he does.

My brother replicates the ballad's opening. He replicates it as closely as can be done on a lone acoustic guitar, the spareness of his sound, the solitariness of it, the difference between the fragility of his version and the orchestra of the original, stirring a longing in the room.

ACKNOWLEDGMENTS

I owe an immeasurable debt to all those who spent long hours with me—and invested much faith in me—as they led me into their lives or into their research. My gratitude begins with everyone who appears in the book but runs well beyond, to all those who shared their experiences or scientific perspectives. "Immeasurable" doesn't begin to describe my debt to my brother, Bob, whose life is this book's inspiration as well as its central story.

Thank you to my agents Sarah Chalfant, Rebecca Nagel, and James Pullen for responding with great care to my ideas, fighting for them, and guiding them into the world.

I'm indebted to my editor, Sarah Murphy, for patient, deeply engaged discussions and for pushing me toward essential solutions. Major thanks to the Ecco team: Helen Atsma, Norma Barksdale, Jin Soo Chun, Meghan Deans, Trina Hunn, Miriam Parker, Allison Saltzman, Rachel Sargent, Paula Szafranski, Lydia Weaver, and Martin Wilson.

Thank you to my keenly insightful editor at the *New York Times Magazine*, Ilena Silverman, with whom I've worked for two decades and who sent me off to the Soteria houses in Israel and to Hearing Voices trainings.

I'm grateful to the New America Foundation, which awarded me a fellowship and an infusion of confidence about this book.

My friends William Hogeland, George Packer, and Laura Secor offered careful readings, crucial encouragement, and sharply focused conversations. They and my friends Paul Barrett, Julie Cohen, Peter Davidson, Samantha Gillison, John Gulla, Ayesha Pande, Saul Shapiro, and Tom Watson have long buoyed my writing and all else.

My parents, amid all the complexities I've recounted, taught me to love and imbued me with what made writing possible.

Maggie and Charlie—I'm lucky to have you in my life.

Georgia—a man could not be luckier!

Natalie and Miles—and Ben and Marli—a father could not be luckier!

And Eden—our languages are different, but your soul is in these pages.

NOTES ON SOURCES

The personal stories I have told are based strictly on extensive interviews, immersion in worlds both shared and unshared, and my own memories. (The book's stories and science are also underpinned by countless interviews with people who do not appear on the page.) The lawyer I call David asked for a minor measure of privacy; his actual name is not David, and I have slightly altered limited details to protect him. The man I have called Ellis requested a name change only. Caroline's clients have also been given pseudonyms.

What follows are citations—chapter by chapter; chronologically within each chapter—for the science and history I write about. The citations do not represent all the tunnels of research that inform the book, but they will provide readers with substantiation and a start to further reading.

Special mention is warranted for the histories of psychiatry contained in two sources, Anne Harrington's *Mind Fixers* and Robert Whitaker's *Mad in America* and *Anatomy of an Epidemic*. Varied criticisms have been made of these two writers, yet it is an understatement to say that the depth of the histories they have compiled is invaluable.

CHAPTER ONE

Caleb Gardner and Arthur Kleinman, "Medicine and the Mind—The Consequences of Psychiatry's Identity Crisis," *New England Journal of Medicine* 381, no. 18 (October 31, 2019): 1697ff.

Ronald Fieve, *Moodswing* (New York: William Morrow, 1975).

For statistics on adult use of psychotropics: Thomas J. Moore and Thomas R. Mattison, "Adult Utilization of Psychiatric Drugs and Differences by Age, Sex, and Race," *Journal of the American Medical Association* (JAMA) Internal Medicine 177, no. 2 (February 2017): 274ff. Also, Debra J. Brody and Qiuping Gu, "Antidepressant Use Among Adults: United States, 2015–2018," *National Center for Health Statistics Data Brief* no. 377 (September 2020). For the forty-fold increase in bipolar diagnosis: Carmen Moreno et al., "National Trends in the Outpatient Diagnosis and Treatment of Bipolar Disorder in Youth," *Archives of General Psychiatry* 64, no. 9 (September 2007): 1032ff. For rates among college students: Marcia R. Morris et al., "Use of Psychiatric Medication by College Students: A Decade of Data," *Pharmacotherapy* 41, no. 4 (April 2021): 350ff.

For ADHD rates of diagnosis: "National Prevalence of ADHD and Treatment: Information on Children and Adolescents, 2016," from the website of the Centers for Disease Control and Prevention, www.cdc.gov/ncbddd /adhd/features/national-prevalence-adhd-and-treatment.html. Also, for further research references and summary of debated ADHD numbers: evolvetreatment.com/blog/diagnosing-adhd-by-country/. Also, Michel Lecendreux et al., "Prevalence of Attention Deficit Hyperactivity Disorder and Associated Features Among Children in France," *Journal of Attention Disorders* 15, no. 6 (August 2011): 516ff. Also, Raphaelle Beau-Lejdstrom et al., "Latest Trends in ADHD Drug Prescribing Patterns in Children in the UK: Prevalence, Incidence, and Persistence," *British Medical Journal* 6 (June 2016).

For the study of antidepressant prescribing patterns in the U.S., UK, and Germany: Carol L. Link et al., "Diagnosis and Management of Depression in Three Countries: Results from a Clinical Vignette Factorial Experiment," *Primary Care Companion for CNS Disorders* 13, no. 5 (2011).

For country-by-country prescription patterns: "Something Startling Is Going on with Antidepressant Use Around the World," *Business Insider*, February 4, 2016.

CHAPTER TWO

On King George: Ida Macalpine and Richard Hunter, *George III and the Mad-Business* (London: Allen Lane, Penguin Press, 1969). Also, the website of the Royal Collection Trust, Georgian Papers Program: www.rct.uk /collection/georgian-papers-programme/medical-papers-relating-to -george-iii, and here: georgianpapersprogramme.com/wp-content/uploads /2018/10/greville.pdf. On Francis Willis's asylum: Frederick Reynolds, *The Life and Times of Frederick Reynolds*, vol. 1 (Philadelphia: W. C. Carey and I. Lea, 1826), 87.

On Benjamin Rush and his tranquilizer chair: Stephen Fried, *Rush: Revolution, Madness, and Benjamin Rush, the Visionary Doctor Who Became a Founding Father* (New York: Crown, 2018), 448. On Benjamin Franklin and electroshock: Tom G. Bolwig and Max Fink, "Electrotherapy for Melancholia: The Pioneering Contributions of Benjamin Franklin and Giovanni Aldini," *Journal of ECT* 25, no. 1 (March 2009): 15ff. Also, Bart Lutters and Peter J. Koehler, "Franklin and Ingenhousz on Cranial Electrotherapy," *World Neurology*, posted March 1, 2016.

On William Cullen: Robert Whitaker, *Mad in America: Bad Science, Bad Medicine, and the Enduring Mistreatment of the Mentally Ill* (New York: Crown, 2002). On the unnamed doctor quoted just after Cullen, see Emil Kraepelin, *One Hundred Years of Psychiatry* (New York: Philosophical Library, 1962); Kraepelin, one of the fathers of modern psychiatry, quotes this doctor only as "Shneider." On Joseph Guislain's Chinese Temple, see his *Traité sur l'Aliénation Mentale et sur les Hospices des Aliénés* vol. 1 (1826). Also, wellcomecollection.org/works/v7t4wd5r. On the story of Guislain's winning entry in the king's competition: museumofthemind.org.uk/projects /european-journeys/bios/joseph-guislain. Also, Whitaker, *Mad in America*.

The story of Phineas Gage and other widely known figures and events in the history of psychiatry and neuroscience will go without citation here. For the quotation from the director of a pastoral refuge, see Whitaker, *Mad in America*. On Nellie Bly's exposé: Nellie Bly, *Ten Days in a Mad-House* (New York: Ian L. Munro, 1887). The story of Henry Cotton is from Whitaker,

Mad in America. The story of the Chicago surgeon—Bayard Holmes—is from Anne Harrington, *Mind Fixers: Psychiatry's Troubled Search for the Biology of Mental Illness* (New York: W. W. Norton & Co., 2019).

On Sigmund Freud: Peter D. Kramer, *Freud: Inventor of the Modern Mind* (New York: HarperCollins, 2006). Also, Mark Solms, "Freud, Luria, and the Clinical Method," *Psychoanalysis and History* 2, no. 1 (2000): 76ff. Also, Sigmund Freud, "Lecture XXXIII: Femininity," *New Introductory Lectures on Psychoanalysis*, 1933.

On the history of the lobotomy: Whitaker, *Mad in America,* and Harrington, *Mind Fixers*. Also, these short films on the procedure and Walter Freeman: www.youtube.com/watch?v=TjPwETslJyc and www.youtube.com/watch?v=Wo2Md95kTCA.

On the risk of tardive dyskinesia with Risperdal: Stephen R. Marder and Tyrone D. Cannon, "Schizophrenia," *New England Journal of Medicine* 381, no. 18 (October 31, 2019): 1753ff. A drug released in 2017, Ingrezza, attempts to treat tardive dyskinesia. Its revenues are around one billion dollars per year. About half of those taking the drug can expect some partial reduction of symptoms, according to research published by the drug's manufacturer, Neurocrine Biosciences, at ingrezzahcp.com.

On the rise of psychoanalysis during and after World War II and on the biological manifesto of 1978: Harrington, *Mind Fixers*.

Donald Goff's analysis in favor of antipsychotics as the default long-term treatment for psychosis: Donald C. Goff et al., "The Long-Term Effects of Antipsychotic Medication on Clinical Course in Schizophrenia," *American Journal of Psychiatry* 174, no. 9 (September 2017): 840ff. A contrary perspective: Lex Wunderlink et al., "Recovery in Remitted First-Episode Psychosis at 7 Years of Follow-up of an Early Dose Reduction/Discontinuation or Maintenance Treatment Strategy: Long-term Follow-up of a 2-year Randomized Clinical Trial," *JAMA Psychiatry* 70, no. 9 (September 2013): 913ff.

CHAPTER FOUR

On Thorazine as well as the minor tranquilizers: Harrington, *Mind Fixers*, and Whitaker, *Mad in America* and *Anatomy of an Epidemic: Magic Bullets, Psychiatric Drugs, and the Astonishing Rise of Mental Illness in America* (New York: Crown, 2010). The Pfizer executive's testimony can be found in the

record of the Senate hearings, p. 10243ff: books.google.com/books?id
=Vt00Zlv38bsC&pg=PA10241&lpg=PA10241&dq=haskell+weinstein
+pfizer&source=bl&ots=i41wXWS_BM&sig=ACfU3U0kF-fDwd3co
BgLYgJvkZ7SmR2nAA&hl=en&sa=X&ved=2ahUKEwij57-987vtAh
WuxVkKHYWuDGYQ6AEwDXoECBAQAg#v=onepage&q=haske
ll%20weinstein%20pfizer&f=false.

The NIMH study of Thorazine and two other antipsychotics: "Phenothiazine
Treatment in Acute Schizophrenia," *Archives of General Psychiatry* 10, no. 3
(1964): 246ff.

On media coverage of Thorazine: "Aid for the Mentally Ill," *New York Times*,
June 26, 1955; "Tranquil Wards Show Drugs' Help," *New York Times*,
January 29, 1956; "Pills for the Mind," *Time*, March 7, 1955.

On President Kennedy: The website of the John F. Kennedy Presidential Library and Museum, www.jfklibrary.org/learn/about-jfk/jfk-in-history
/john-f-kennedy-and-people-with-intellectual-disabilities. Also, "President
Seeks Funds to Reduce Mental Illness," *New York Times*, February 6, 1963.

On the development of Haldol: Bernard Granger and Simona Albu, "The
Haloperidol Story," *Annals of Clinical Psychiatry* 17, no. 1 (2005): 137ff.

On antipsychotic dosages: S. P. Segal et al., "Neuroleptic Medication and
Prescription Practices with Sheltered-Care Residents: A 12-year Perspec-
tive," *American Journal of Public Health* 82, no. 6 (June1992): 846ff. Also,
G. T. Reardon et al., "Changing Patterns of Neuroleptic Dosage Over a
Decade," *American Journal of Psychiatry* 146, no. 6 (June 1989): 726ff.

On the Oprah interview and fallout: "Free Expression or Irresponsibility,"
New York Times, September 22, 1987.

The Pulitzer-winning series on psychotropic medication: Jon Franklin,
"The Mind Fixers," Baltimore *Sun*, July 23–31, 1984.

LSD's role: Harrington, *Mind Fixers*.

NAMI's IRS filing: Whitaker, *Anatomy of an Epidemic*.

On Zyprexa: "Eli Lilly Said to Play Down Risks of Top Pill," *New York Times*,
December 17, 2006. Also, "U.S. Drug Agency Investigating Accuracy of
Lilly's Zyprexa Data," *New York Times*, April 25, 2007. Also, "Lilly Said
to Be Near $1.4 Billion U.S. Settlement," *New York Times*, January 14,
2009. Also, Jim Gottstein, *The Zyprexa Papers* (Anchorage: Jim Gottstein,
2020). On Zyprexa and foster children: Thomas I. Mackie et al., "Antipsychotic Use Among Youth in Foster Care Enrolled in a Specialized
Managed Care Organization Intervention," *Journal of the American*

Academy of Child and Adolescent Psychiatry 59, no. 1 (January 2020): 166ff. Also, Meredith Matone et al., "Antipsychotic Prescribing to Children: An In-Depth Look at Foster Care and Medicaid Populations," Policy Lab, The Children's Hospital of Philadelphia, Spring 2015. Also, "Why Are So Many Foster Care Children Taking Antipsychotics?" *Time*, November 29, 2011.

On Joseph Biederman: Whitaker, *Anatomy of an Epidemic*.

The account of the marketing and side effects of Risperdal owes a great debt to Steven Brill's series "America's Most Admired Lawbreaker," *Huffington Post*, 2015.

The NIMH study of second-generation antipsychotics (stating, "Neurologic Side Effects: There were no significant differences among the groups in the incidence of extrapyramidal side effects, akathisia, or movement disorders as reflected by rating-scale measures of severity"): Jeffrey A. Lieberman et al., "Effectiveness of Antipsychotic Drugs in Patients with Chronic Schizophrenia," *New England Journal of Medicine* 353, no. 12 (September 22, 2005): 1209ff.

CHAPTER FIVE

David's reading about benzodiazepine withdrawal: Baylissa Frederick, *Recovery and Renewal: Your Essential Guide to Overcoming Dependency and Withdrawal from Sleeping Pills, Other Benzodiazepine Tranquillisers, and Antidepressants* (RRW Publishing, 2017). Also, Heather Ashton, *Benzodiazepines: How They Work and How to Withdraw*, online at www.benzo.org.uk/manual/bzcha01.htm.

The Nathan Kline quotation is as quoted in Harrington, *Mind Fixers*.

The theory of deficient neurotransmitters: J. J. Schildkraut, "The Catecholamine Hypothesis of Affective Disorders: A Review of Supporting Evidence," *American Journal of Psychiatry* 122, no. 5 (November 1965): 509ff.

An example of the chemical-imbalance ads: www.adforum.com/creative-work/ad/player/23659/christmas-tree/prozac. On the marketing of antidepressants: Nathan P. Greenslit and Ted J. Kaptchuk, "Antidepressants and Advertising: Psychopharmaceuticals in Crisis," *Yale Journal of Biology and Medicine* 85, no. 1 (March 2012): 153ff.

The research on serotonin levels is as described and cited in Whitaker, *Anatomy of an Epidemic*, except for the University of Texas research: P. L. Delgado and F. A. Moreno, "Role of Norepinephrine in Depression," *Journal of Clinical Psychiatry* 61, suppl. 1 (2000): 5ff.

The 2021 NIMH language about chemical imbalance: www.ncbi.nlm.nih
.gov/books/NBK361016/#:~:text=In%20other%20words%2C
%20antidepressants%20improved,within%20one%20or%20two%20weeks.

On the placebo effect: Irving Kirsch, *The Emperor's New Drugs: Exploding the
Antidepressant Myth* (New York: Random House, 2009). Peter D. Kramer,
Ordinarily Well: The Case for Antidepressants (New York: Farrar, Straus and
Giroux, 2016). Also, Marcia Angell, "The Epidemic of Mental Illness:
Why?, *New York Review of Books,* June 23, 2011. On the rise in use of anti-
depressants 2009–2018: Debra J. Brody and Qiuping Gu, "Antidepressant
Use Among Adults: United States, 2015–2018," *National Center for Health
Statistics Data Brief* no. 377 (September 2020).

On McKinsey & Company: "McKinsey Advised Purdue Pharma How to
'Turbocharge' Opioid Sales, Lawsuit Says," *New York Times*, February 1,
2019.

On antidepressant withdrawal: "Many People Taking Antidepressants Discover They Cannot Quit," *New York Times*, April 7, 2018. Also, "How to
Quit Antidepressants: Very Slowly, Doctors Say," *New York Times*, March
5, 2019. Also, Rachel Aviv, "The Challenge of Going Off Psychiatric
Drugs," the *New Yorker*, April 1, 2019. Also, Abie Horowitz and David
Taylor, "Tapering of SSRI Treatment to Mitigate Withdrawal Symptoms," *The Lancet Psychiatry* 6, no. 6 (June 2019): 538ff.

The 1997 report, produced by Eli Lilly, on "discontinuation syndrome":
www.baumhedlundlaw.com/documents/pdf/cymbalta-unsealed-docs
/Exh-078-Journal-Clinical-Psychiatry-Supplement-Antidepressant
-Discontinuation-Syndrome-Update-SSRIs-1997.pdf.

CHAPTER SIX

The interrogation of Joan of Arc: www.jeanne-darc.info/trials-index/the
-examination-at-poitiers/.

CHAPTER EIGHT

On connectivity: Esther M. Blessing et al., "Anterior Hippocampal–Cortical
Functional Connectivity Distinguishes Antipsychotic Naïve First-Episode
Psychosis Patients from Controls and May Predict Response to Second-
Generation Antipsychotic Treatment," *Schizophrenia Bulletin* 46, no. 3
(May 2020): 680ff.

On the learning and unlearning of fear: Daphne J. Holt et al., "Extinction Memory Is Impaired in Schizophrenia," *Biological Psychiatry* 65, no. 6 (March 2009): 455ff.

Goff's journey with DCS, a sampling of papers: Donald C. Goff et al., "Dose-Finding Trial of D-cyloserine Added to Neuroleptics for Negative Symptoms in Schizophrenia," *American Journal of Psychiatry* 152, no. 8 (August 1995): 1213ff. Also, Donald C. Goff and Joseph T. Coyle, "The Emerging Role of Glutamate in the Pathophysiology and Treatment of Schizophrenia," *American Journal of Psychiatry* 158, no. 9 (September 2001): 1367ff. Also, Donald C. Goff, "D-Cycloserine: An Evolving Role in Learning and Neuroplasticity in Schizophrenia," *Schizophrenia Bulletin* 38, no. 5 (September 2012): 936ff. Also, Erica D. Diminich, "D-cycloserine Augmentation of Cognitive Behavioral Therapy for Delusions: A Randomized Clinical Trial," *Schizophrenia Research* 222 (August 2020): 145ff.

On possible causes of schizophrenia: Jeffrey A. Lieberman et al., eds., *The American Psychiatric Association Publishing Textbook of Schizophrenia* (Washington, D.C.: American Psychiatric Association Publishing, 2020). The Lieberman textbook is one source for the six percent chance of being diagnosed with schizophrenia if one biological parent has the condition. Also on possible causes, Rebecca Pinto et al., "Schizophrenia in Black Caribbeans Living in the U.K.: An Exploration of Underlying Causes of the High Incidence Rate," *British Journal of General Practice* 58, no. 551 (June 2008): 429ff.

CHAPTER TEN

On Soteria's history: Loren R. Mosher and Voyce Hendrix, *Soteria: Through Madness to Deliverance* (Loren R. Mosher, Voyce Hendrix, 2004).

On the WHO research: Assen Jablensky et al., "Schizophrenia: Manifestations, Incidence, and Course in Different Cultures," *Psychological Medicine Monograph Supplement* 20 (1992). On the critiques: Alex Cohen et al., "Questioning an Axiom: Better Prognosis for Schizophrenia in the Developing World?" *Schizophrenia Bulletin* 34, no. 2 (March 2008): 229ff. And a counter to the critiques: Assen Jablensky and Norman Sartorius, "What Did the WHO Studies Really Find?" *Schizophrenia Bulletin* 34, no. 2 (March 2008): 253ff.

The Dutch research: Lex Wunderlink et al., "Recovery in Remitted First-Episode Psychosis at 7 Years of Follow-up of an Early Dose Reduction /Discontinuation or Maintenance Treatment Strategy: Long-term Follow-up of a 2-year Randomized Clinical Trial," *JAMA Psychiatry* 70, no. 9 (September 2013): 913ff.

On brain atrophy: "Using Images to Understand How the Brain Works," *New York Times*, September 16, 2008.

On schizophrenia and violence: Harry J. Steadman et al., "Violence by People Discharged from Acute Psychiatric Inpatient Facilities and by Others in the Same Neighborhoods," *Archives of General Psychiatry* 55, no. 5 (May 1998): 393ff. Also, Seena Fazel et al., "Schizophrenia and Violence: Systematic Review and Meta-analysis," *PLOS Medicine* 6, no. 8 (August 2009). Also, Paul S. Appelbaum, "Violent Acts and Being the Target of Violence Among People With Mental Illness—the Data and Their Limits," *JAMA Psychiatry* 77, no. 4 (April 2020): 345ff. Also, Daniel Whiting et al., "Violence and Mental Disorders: A Structured Review of Associations by Individual Diagnoses, Risk Factors, and Risk Assessment," *Lancet Psychiatry* 8 (February 2021): 150ff.

The OnTrackNY manual: www.ontrackny.org/portals/1/Files/Resources /MedicalManual_2015.01.21.pdf.

The coverage of psychedelics as medication: "The Psychedelic Revolution Is Coming. Psychiatry May Never Be the Same," *New York Times*, May 5, 2021. On the trial with no significant difference: Robin Carhart-Harris, "Trial of Psilocybin Versus Escitalopram for Depression," *New England Journal of Medicine* 384, no. 15 (April 15, 2021): 1402ff.

CHAPTER ELEVEN

The writing of Marius Romme: Marius A. Romme and Alexandra D. Escher, "Hearing Voices," *Schizophrenia Bulletin* 15, no. 2 (1989): 209ff.